U.S.NRC

United States Nuclear Regulatory Commission

Protecting People and the Environment

NUREG-1307, Rev. 15

Report on Waste Burial Charges

Changes in Decommissioning Waste Disposal Costs at Low-Level Waste Burial Facilities

Final Report

Office of Nuclear Reactor Regulation

AVAILABILITY OF REFERENCE MATERIALS
IN NRC PUBLICATIONS

NUREG-1307, Rev. 15

United States Nuclear Regulatory Commission

Protecting People and the Environment

Report on Waste Burial Charges

Changes in Decommissioning Waste Disposal Costs at Low-Level Waste Burial Facilities

Final Report

Manuscript Completed: December 2012
Date Published: January 2013

Prepared by:
Pacific Northwest National Laboratory
Jason A. Gastelum and Steven Short

NRC Project Manager:
Jo Ann Simpson

Office of Nuclear Reactor Regulation

Disclaimer

This work report was prepared as an account of work sponsored by an agency of the U.S. Government. Neither the U.S. Government nor any agency thereof, or any of their employees, make any warranty, expressed or implied, or assumes any legal liability or responsibility for any third party's use or the results of such use, of any information, apparatus, product, or process disclosed in this report, or represents that its use by such third party would not infringe on privately owned rights. The views expressed in this paper are not necessarily those of the U.S. Nuclear Regulatory Commission (NRC).

NUREG-1307, Revision 15, is not a substitute for NRC regulations. The approaches and methods described in this NUREG are provided for information only. Publication of this report does not necessarily constitute NRC approval or agreement with the information contained herein.

ABSTRACT

The U.S. Nuclear Regulatory Commission (NRC) requires nuclear power reactor licensees to annually adjust the estimate of the cost of decommissioning their plants, in dollars of the current year, as part of the process providing reasonable assurance that adequate funds for decommissioning will be available when needed. This report, which is revised periodically, explains the formula acceptable to the NRC for determining the minimum decommissioning fund requirements for nuclear power plants. The sources of information used in the formula are identified, and the values developed for the estimation of radioactive waste burial/disposition costs, by site and year, are given. Licensees may use the formula, coefficients, and burial/disposition adjustment factors from this report in their cost analyses, or they may use adjustment factors derived from any methodology that results in a total cost estimate of no less than the amount estimated by using the parameters presented in this report.

This 15th revision of NUREG-1307 contains disposal costs updated to the year 2012 for the reference pressurized-water reactor (PWR) and the reference boiling-water reactor (BWR). Two different options for estimating these costs are presented. The first option assumes that 100% of the low-level waste (LLW) generated during decommissioning is disposed of at one of the two compact-affiliated disposal facilities, located in Richland, Washington, and Barnwell, South Carolina. Ratios of year 2012 disposal costs to the original year 1986 disposal costs (i.e., B_x factors) are also provided. For historical purposes, disposal costs for the reference reactors and ratios of disposal costs at the Washington and South Carolina sites for the years 2002, 2004, 2006, 2008, and 2010 are provided.

The second option provides for disposing of LLW using a combination of non-compact and compact-affiliated disposal facilities. This option, which is accepted as a rational alternative for licensees to pursue, allows nuclear power plant (NPP) licensees to take advantage of potentially lower disposal costs for much of their decommissioning LLW. As with the first option, ratios of the alternative disposal costs to the original year 1986 disposal costs (i.e., B_x factors) are provided.

Several sample calculations for estimating the burial/disposition cost for both options are presented, demonstrating the use of the data contained in this report.

Estimated disposal costs for 2012 using only the Washington disposal site, which accepts LLW from members of the Northwest and Rocky Mountain Compacts, are about 9 percent lower for the reference PWR and 10 percent lower for the reference BWR when compared to 2010 costs. The decreases in disposal costs were driven by decreases in volume and container charges at the Washington disposal site. Disposal costs for the option in which a portion of decommissioning LLW is disposed of at a non-compact disposal facility are about 12 percent higher for the PWR and 11 percent higher for the BWR when compared to 2010 costs. The increases are due to increases in the non-compact disposal facility rates.

Estimated disposal costs for 2012 using only the South Carolina disposal site, which accepts LLW from members of the Atlantic Compact, are about 12 percent higher for both the reference PWR and reference BWR when compared to 2010 costs. The increases in disposal costs were driven by increases in weight, curie, and irradiated hardware charges at the South Carolina site. Disposal costs for the option in which a portion of the decommissioning LLW is disposed of at a non-compact disposal facility are about 13 percent higher for both the PWR and BWR when compared to 2010 costs. The increases are due to increases in the non-compact disposal facility rates.

Currently, NUREG-1307, Revision 15, assumes that LLW generated during plant operations is disposed of using operating funds. Plants that have no disposal site available for LLW are now forced to provide interim storage for this waste (although most Class A waste can be disposed of at the non-compact disposal facility located in Clive, Utah). If additional disposal sites do not become available before permanent plant shutdown, this waste ultimately will need to be disposed of during decommissioning. This volume can become significant for plants operating through extended license terms, and the disposal cost would not be accounted for in a decommissioning trust fund based on the formula calculation.

For plants that have no disposal site available for Class B and C LLW (e.g., plants not located within the Atlantic and Northwest Compacts), NUREG-1307, Revision 15, assumes that the cost for disposal of this waste is the same as that for the Atlantic Compact, for lack of a better alternative at this time. As new disposal options become available, they will be incorporated into subsequent revisions of NUREG-1307.

FOREWORD

Nuclear power reactor licensees are required by Title 10 of the *Code of Federal Regulations* (10 CFR) 50.75, "Reporting and Recordkeeping for Decommissioning Planning," to annually adjust the estimated decommissioning costs of their nuclear facilities to ensure adequate funds are available for decommissioning. The regulation references NUREG-1307 as the appropriate source for obtaining the adjustment factor for waste burial/disposition costs. Revision 15 of NUREG-1307 provides current waste burial/disposition costs using the compact-affiliated disposal facilities located in Richland, Washington, and Barnwell, South Carolina. In addition, this revision also includes an alternative disposal cost that provides the option of disposing of LLW using a combination non-compact and compact-affiliated disposal facilities. Licensees can factor these numbers into the adjustment formula, as specified in 10 CFR 50.75(c)(2), to determine the minimum decommissioning fund requirement for their nuclear facilities. Although this report is specifically prepared for the use of power reactor licensees, it also can be a valuable source of information for material licensees on current waste burial/disposition costs.

On July 1, 2000, the Barnwell disposal facility at Barnwell, South Carolina, became the host disposal facility for the newly formed Atlantic Compact, comprised of the States of Connecticut, New Jersey, and South Carolina. Effective July 1, 2008, LLW from States that are not members of the Atlantic Compact was no longer accepted at the South Carolina disposal site. The South Carolina Public Service Commission annually determines the costs of waste disposal at the Barnwell disposal facility and provides the site operator with an allowable operating margin. The Richland, Washington, facility only accepts LLW from the Northwest and Rocky Mountain Compacts. The costs of disposal for this facility are determined annually based on waste generator volume projections and a maximum annual operator revenue set by the Washington Utilities and Transportation Commission. If the total operator revenue is exceeded in a given year, a rebate may be sent to the waste generator.

Since the Barnwell and Richland LLW disposal facilities are only available to nuclear facilities located within the respective compacts, another option available to licensees is to dispose of decommissioning Class A LLW at a non-compact disposal facility. Costs for this option are based on a price quote received from the operator of the non-compact disposal facility located in Clive, Utah. This report provides waste burial/disposition adjustment factors for this option, in addition to the standard option of disposing of 100% of decommissioning LLW at a compact-affiliated disposal facility.

Low-level radioactive waste disposal costs are an important element in the cost of decommissioning a nuclear facility. This report provides the latest information available at the time of publication for licensees to use for annually adjusting the estimated cost of decommissioning their nuclear facilities.

Currently, NUREG-1307, Revision 15, assumes that LLW generated during plant operations is disposed of using operating funds. Plants that have no disposal site available for LLW are now forced to provide interim storage for this waste (although most Class A waste can be disposed of at the non-compact disposal facility located in Clive, Utah). If additional disposal sites do not become available before permanent plant shutdown, this waste ultimately will need to be disposed of during decommissioning. This volume can become significant for plants operating through extended license terms, and the disposal cost would not be accounted for in a decommissioning trust fund based on the formula calculation. In addition, for plants that have no disposal site available for LLW (e.g., plants not located within the Atlantic, Northwest, and

Rocky Mountain Compacts), NUREG–1307, Revision 15, assumes the cost for disposal is the same as that provided for the Atlantic Compact, for lack of a better alternative at this time. However, when new disposal facilities become available, disposal rates likely will be significantly higher. Accordingly, given these considerations, licensees may want to set aside additional decommissioning trust funds to avoid significant future shortfalls in funding and potential enforcement actions.

For plants that have no disposal site available for Class B and C LLW (e.g., plants not located within the Atlantic and Northwest Compacts), NUREG-1307, Revision 15, assumes that the cost for disposal of this waste is the same as that for the Atlantic Compact, for lack of a better alternative at this time. As new disposal options become available, they will be incorporated into subsequent revisions of NUREG-1307.

Ho K. Nieh, Director
Division of Inspection and Regional Support
Office of Nuclear Reactor Regulation

TABLE OF CONTENTS

Section   **Page**

ABSTRACT...iii
FOREWORD v
1.   **INTRODUCTION**...1
2.   **SUMMARY**..3
3.   **DEVELOPMENT OF COST ADJUSTMENT FORMULA**...5
     3.1   Labor Adjustment Factors 7
     3.2   Energy Adjustment Factors 7
     3.3   Waste Burial Adjustment Factors 8
     3.4   Sample Calculations of Estimated Reactor Decommissioning Costs..................8
4.   **REFERENCES**..11
APPENDIX A.   LOW-LEVEL WASTE BURIAL/DISPOSITION PRICES FOR THE
                        CURRENT YEAR A-1
APPENDIX B.   CALCULATION OF LOW-LEVEL WASTE BURIAL/DISPOSITION
                        COST ESTIMATION FACTORS...B-1
APPENDIX C.   BUREAU OF LABOR STATISTICS ON THE INTERNET C-1
APPENDIX D.   REPRESENTATIVE EXAMPLES OF DECOMMISSIONING COSTS
                        FOR 2002 THROUGH 2012 ....D-1
APPENDIX E.   LOW-LEVEL WASTE COMPACTS E-1

LIST OF TABLES

Table 2-1   Values of B_x as a Function of LLW Burial Site, Waste Vendor, and Year 4

Table 3-1   Evaluation of the Coefficients A, B, and C in January 1986 Dollars.................... 6

Table 3-2   Regional Factors for Labor Cost Adjustment....................................... 7

Table A-1   Schedule of Maximum Allowable LLW Disposal at the South Carolina
Disposal Facility A-2

Table A-2   Price Quotes for Disposition of Class A LLW at the Non-Compact Disposal
Facility Located in Clive, Utah...A-4

Table B-1   PWR Burial Costs at the Washington Site (2012 dollars) B-3

Table B-2   BWR Burial Costs at the Washington Site (2012 dollars) B-4

Table B-3   PWR Burial Costs at the Washington Site (2010 dollars) B-5

Table B-4   BWR Burial Costs at the Washington Site (2010 dollars) B-6

Table B-5   PWR Burial Costs at the Washington Site (2008 dollars) B-7

Table B-6   BWR Burial Costs at the Washington Site (2008 dollars) B-8

Table B-7   PWR Burial Costs at the Washington Site (2006 dollars) B-9

Table B-8   BWR Burial Costs at the Washington Site (2006 dollars) B-10

Table B-9   PWR Burial Costs at the Washington Site (2004 dollars) B-11

Table B-10   BWR Burial Costs at the Washington Site (2004 dollars) B-12

Table B-11   PWR Burial Costs at the Washington Site (2002 dollars) B-13

Table B-12   BWR Burial Costs at the Washington Site (2002 dollars) B-14

Table B-13   PWR Burial Costs at the South Carolina Site Atlantic Compact
(2012 dollars)...B-15

Table B-14   BWR Burial Costs at the South Carolina Site Atlantic Compact
(2012 dollars)...B-16

Table B-15   PWR Burial Costs at the South Carolina Site Atlantic Compact
(2010 dollars)...B-17

Table B-16   BWR Burial Costs at the South Carolina Site Atlantic Compact
(2010 dollars)...B-18

Table B-17   PWR Burial Costs at the South Carolina Site Atlantic Compact
(2008 dollars)...B-19

Table B-18   BWR Burial Costs at the South Carolina Site Atlantic Compact
(2008 dollars)...B-20

Table B-19   PWR Burial Costs at the South Carolina Site Atlantic Compact
(2006 dollars)...B-21

Table B-20   BWR Burial Costs at the South Carolina Site Atlantic Compact
(2006 dollars)...B-22

Table B-21   PWR Burial Costs at the South Carolina Site Non-Atlantic Compact
(2006 dollars)...B-23

Table B-22   BWR Burial Costs at the South Carolina Site Non-Atlantic Compact
(2006 dollars)...B-24

Table B-23   PWR Burial Costs at the South Carolina Site Atlantic Compact
(2004 dollars)...B-25

Table B-24   BWR Burial Costs at the South Carolina Site Atlantic Compact
(2004 dollars)...B-26

Table B-25   PWR Burial Costs at the South Carolina Site Non-Atlantic Compact
(2004 dollars)..B-27

Table B-26   BWR Burial Costs at the South Carolina Site Non-Atlantic Compact
(2004 dollars)..B-28

Table B-27   PWR Burial Costs at the South Carolina Site Atlantic Compact
(2002 dollars)..B-29

Table B-28   BWR Burial Costs at the South Carolina Site Atlantic Compact
(2002 dollars)..B-30

Table B-29   BWR Burial Costs at the South Carolina Site Non-Atlantic Compact
(2002 dollars)..B-31

Table B-30   BWR Burial Costs at the South Carolina Site Non-Atlantic Compact
(2002 dollars)..B-32

Table B-31   PWR LLW Disposition Costs Using a Combination of Non-Compact Disposal
Facility and the Washington Disposal Facility (2012 dollars)..........................B-33

Table B-32   BWR LLW Disposition Costs Using a Combination of Non-Compact Disposal
Facility and the Washington Disposal Facility (2012 dollars)..........................B-34

Table B-33   PWR LLW Disposition Costs Using a Combination of Non-Compact Disposal
Facility and the South Carolina Disposal Facility (2012 dollars) B-35

Table B-34   BWR LLW Disposition Costs Using a Combination of Non-Compact Disposal
Facility and the South Carolina Disposal Facility (2012 dollars) B-36

Table B-35   PWR Disposition Costs Using Waste Vendors with Burial Costs at the
Washington Site (2010 dollars) B-37

Table B-36   BWR Disposition Costs Using Waste Vendors with Burial Costs at the
Washington Site (2010 dollars) B-38

Table B-37   PWR Disposition Costs Using Waste Vendors with Burial Costs at the South
Carolina Site (2010 dollars)...B-39

Table B-38   BWR Disposition Costs Using Waste Vendors with Burial Costs at the South
Carolina Site (2010 dollars)...B-40

Table B-39   PWR Disposition Costs Using Waste Vendors with Burial Costs at the
Washington Site (2008 dollars) B-41

Table B-40   BWR Disposition Costs Using Waste Vendors with Burial Costs at the
Washington Site (2008 dollars) B-42

Table B-41   PWR Disposition Costs Using Waste Vendors with Burial Costs at the
South Carolina Site Atlantic Compact (2008 dollars).....................................B-43

Table B-42   BWR Disposition Costs Using Waste Vendors with Burial Costs at the
South Carolina Site Atlantic Compact (2008 dollars).....................................B-44

Table B-43   PWR Disposition Costs Using Waste Vendors with Burial Costs at the
Washington Site (2006 dollars) B-45

Table B-44   BWR Disposition Costs Using Waste Vendors with Burial Costs at the
Washington Site (2006 dollars) B-46

Table B-45   PWR Disposition Costs Using Waste Vendors with Burial Costs at the
South Carolina Site Atlantic Compact (2006 dollars).....................................B-47

Table B-46   BWR Disposition Costs Using Waste Vendors with Burial Costs at the
South Carolina Site Atlantic Compact (2006 dollars).....................................B-48

Table B-47   PWR Disposition Costs Using Waste Vendors with Burial Costs at the
South Carolina Site Non-Atlantic Compact (2006 dollars) B-49

Table B-48   BWR Disposition Costs Using Waste Vendors with Burial Costs at the
              South Carolina Site Non-Atlantic Compact (2006 dollars) B-50
Table B-49   PWR Disposition Costs Using Waste Vendors with Burial Costs at the
              Washington Site (2004 dollars) B-51
Table B-50   BWR Disposition Costs Using Waste Vendors with Burial Costs at the
              Washington Site (2004 dollars) B-52
Table B-51   PWR Disposition Costs Using Waste Vendors with Burial Costs at the
              South Carolina Site Atlantic Compact (2004 dollars) B-53
Table B-52   BWR Disposition Costs Using Waste Vendors with Burial Costs at the
              South Carolina Site Atlantic Compact (2004 dollars) B-54
Table B-53   PWR Disposition Costs Using Waste Vendors with Burial Costs at the
              South Carolina Site Non-Atlantic Compact (2004 dollars) B-55
Table B-54   BWR Disposition Costs Using Waste Vendors with Burial Costs at the
              South Carolina Site Non-Atlantic Compact (2004 dollars) B-56
Table B-55   PWR Disposition Costs Using Waste Vendors with Burial Costs at the
              Washington Site (2002 dollars) B-57
Table B-56   BWR Disposition Costs Using Waste Vendors with Burial Costs at the
              Washington Site (2002 dollars) B-58
Table B-57   PWR Disposition Costs Using Waste Vendors with Burial Costs at the
              South Carolina Site Atlantic Compact (2002 dollars) B-59
Table B-58   BWR Disposition Costs Using Waste Vendors with Burial Costs at the
              South Carolina Site Atlantic Compact (2002 dollars) B-60
Table B-59   PWR Disposition Costs Using Waste Vendors with Burial Costs at the
              South Carolina Site Non-Atlantic Compact (2002 dollars) B-61
Table B-60   BWR Disposition Costs Using Waste Vendors with Burial Costs at the
              South Carolina Site Non-Atlantic Compact (2002 dollars) B-62

1. INTRODUCTION

Guidance in Title 10 of the *Code of Federal Regulations* (10 CFR) 50.75(b) States that the U.S. Nuclear Regulatory Commission (NRC) requires nuclear power plant licensees to annually adjust the estimate of the cost (in dollars of the current year) of decommissioning their plants. This is one step of a multi-step process of providing reasonable assurance to the NRC that adequate funds for decommissioning will be available when needed. This report provides adjustment factors for the waste burial/disposition component of the decommissioning fund requirement, as required by 10 CFR 50.75(c)(2). This report also provides the regional adjustment factors for the labor and energy components of the decommissioning fund requirement. The term "adjustment factor," as used in this report and in 10 CFR 50.75(c)(2), refers to increases and decreases in decommissioning costs since the NRC regulations were issued. The decommissioning fund requirements in these regulations are in 1986 dollars. This report is updated periodically to reflect changes in waste burial/disposition costs.

This report provides the development of a formula for estimating decommissioning cost that is acceptable to the NRC. Sources of information used in the formula are identified. Values developed for the adjustment of radioactive waste burial/disposition costs, by site and by year, are also given. Licensees may use the formula, the coefficients, and the burial/disposition adjustment factors from this report in their analyses, or they may use an adjustment factor at least equal to the approach presented herein.

The formula and its coefficients, together with guidance to the appropriate sources of data needed, are summarized in Chapter 2. The development of the formula and its coefficients, with sample calculations, is presented in Chapter 3. Price schedules for burial/disposition for the year 2012 are given in APPENDIX A for compact-affiliated and non-compact disposal facilities. Calculations to determine the burial/disposition adjustment factors, B_x, for each site and year of evaluation are summarized in 4APPENDIX B.

This 15th revision of NUREG-1307 contains disposal costs updated to the year 2012 for the reference pressurized-water reactor (PWR) and the reference boiling-water reactor (BWR). Two different options for estimating these costs are presented. The first option assumes that 100% of the low-level waste (LLW) generated during decommissioning is disposed of at one of the two compact-affiliated disposal facilities located, in Richland, Washington and Barnwell, South Carolina. Ratios of 2012 disposal costs to the original 1986 disposal costs (i.e., B_x factors) are also provided. For historical purposes, disposal costs for the reference reactors and ratios of disposal costs at the Washington and South Carolina sites for the years 2002, 2004, 2006, 2008, and 2010 are provided.

The second option provides for disposing of LLW using a combination of non-compact and compact-affiliated disposal facilities. For a PWR under this option, 93% of the LLW is assumed to be disposed of at a non-compact disposal facility and the remaining 7% is assumed to be disposed of at a compact-affiliated disposal facility. For a BWR under this option, 95% of the LLW is assumed to be disposed of at a non-compact disposal facility and the remaining 5% is assumed to be disposed of at a compact-affiliated disposal facility. This option, which is accepted as a rational alternative for licensees to pursue, allows NPP licensees to take advantage of potentially lower disposal costs for much of their LLW waste. Ratios of the 2012 alternative disposal costs to the original year 1986 disposal costs (i.e., B_x factors) are also provided.

Currently, NUREG-1307, Revision 15, assumes that LLW generated during plant operations is disposed of using operating funds. Plants that have no disposal site available for LLW are now forced to provide interim storage for this waste (although most Class A waste can be disposed of at the non-compact disposal facility located in Clive, Utah). If additional disposal sites do not become available before permanent plant shutdown, this waste ultimately will need to be disposed of during decommissioning. This volume can become significant for plants operating through extended license terms, and the disposal cost would not be accounted for in a decommissioning trust fund based on the formula calculation.

For plants that have no disposal site available for Class B and C LLW (e.g., plants not located within the Atlantic and Northwest Compacts), NUREG-1307, Revision 15, assumes that the cost for disposal of this waste is the same as that for the Atlantic Compact, for lack of a better alternative at this time. As new disposal options become available, they will be incorporated into subsequent revisions of NUREG-1307.

2. SUMMARY

The elements of decommissioning cost, per Title 10 of the *Code of Federal Regulations* (10 CFR) 50.75(c)(2)), are assigned to three categories: those that are proportional to labor costs, L_x, those that are proportional to energy costs, E_x, and those that are proportional to burial costs, B_x. The adjustment of the total decommissioning cost estimate can be expressed by:

$$\text{Estimated cost (Year X)} = [1986 \text{ \$ cost}] [A*L_x + B*E_x + C*B_x]$$

where A, B, and C are the fractions of the total 1986 dollar costs attributable to labor (0.65), energy (0.13), and burial (0.22), respectively, and sum to 1.0. The factors L_x, E_x, and B_x are defined by:

L_x = labor cost adjustment, January of 1986 to the latest month of Year X for which data are available,

E_x = energy cost adjustment, January of 1986 to the latest month of Year X for which data are available, and,

B_x = LLW burial/disposition cost adjustment, January of 1986 to the latest month of Year X for which data are available.

Licensees are to evaluate L_x and E_x for the years subsequent to 1986 based on the national producer price indexes, national consumer price indexes, and local conditions for a given site (see Chapter 3).

B_x is evaluated by recalculating the costs of burial/disposition of the radioactive wastes from the reference PWR (Ref. 1) and the reference BWR (Ref. 2) based on the price schedules provided by the available disposal facilities for the year of interest. The results of these recalculations are presented in Table 2-1, by site and by year. Effective January 1, 1993, waste from States that are not members of the Northwest or Rocky Mountain Compacts was no longer accepted at the Washington disposal site. Effective July 1, 2000, different price schedules at the South Carolina burial site applied for States within and outside the newly created Atlantic Compact, comprised of South Carolina, Connecticut, and New Jersey (see footnote (c) in Table 2-1). Effective July 1, 2008, waste from States that are not members of the Atlantic Compact was no longer accepted at the South Carolina disposal site. Licensees not located in either the Northwest, Rocky Mountain, or Atlantic Compacts should use the B_x values for the Generic LLW Disposal Site (see footnote (e) in Table 2-1). Effective with Revision 8, the additional option of using a combination of waste vendors, or non-compact disposal facilities, and compact-affiliated disposal facilities was made available, and was referred to as "Direct Disposal with Vendors". Effective with Revision 15, the nomenclature for the two disposal options, referred to as "Direct Disposal" and "Direct Disposal with Vendors" in previous revisions of NUREG-1307, is changed to "Compact-Affiliated Disposal Facility Only" and "Combination of Compact-Affiliated and Non-Compact Disposal Facilities" to better describe these options. The B_x values for this option are also provided in Table 2-1 (see footnotes (f) and (g) in Table 2-1). The decision rests with the licensees to determine the option that best represents their particular situation.

Table 2-1. Values of B_x as a Function of LLW Burial Site, Waste Vendor, and Year[a]

	B_x Values for Washington Site[b]				B_x Values for South Carolina Site								B_x Values for Generic LLW Disposal Site[e]			
					Atlantic Compact[c]				Non-Atlantic Compact[d]							
	Compact-Affiliated Facility Only[g]		Combination of Compact-Affiliated and Non-Compact Facility[f,g]		Compact-Affiliated Facility Only[g]		Combination of Compact-Affiliated and Non-Compact Facility[f,g]		Compact-Affiliated Facility Only[g]		Combination of Compact-Affiliated and Non-Compact Facility[f,g]		Compact-Affiliated Facility Only[g]		Combination of Compact-Affiliated and Non-Compact Facility[f,g]	
Year	PWR	BWR	PWR	BWR	PWR	BWR	PWR	BWR	PWR	BWR	PWR	BWR	PWR	BWR	PWR	BWR
2012	7.335	6.704	7.375	6.076	30.581	27.295	13.885	14.160	NA	NA	NA	NA	30.581	27.295	13.885	14.160
2010	8.035	7.423	6.588	5.458	27.292	24.356	12.280	12.540	NA	NA	NA	NA	27.292	24.356	12.280	12.540
2008	8.283	23.185	5.153	20.889	25.231	22.504	9.872	11.198	NA	NA	NA	NA	25.231	22.504	9.872	11.198
2006	6.829	11.702	3.855	9.008	22.933	20.451	8.600	9.345	23.030	20.813	8.683	10.206	NA	NA	NA	NA
2004	5.374	13.157	3.846	11.755	19.500	17.389	7.790	8.347	21.937	17.970	7.934	8.863	NA	NA	NA	NA
2002	3.634	14.549	5.748	15.571	17.922	15.988	9.273	8.626	18.732	16.705	9.467	8.860	NA	NA	NA	NA

(a) The values shown in this table are developed in APPENDIX B, with all values normalized to the 1986 Washington PWR and BWR values by dividing the calculated burial costs for each site and year by the Washington site burial costs calculated for the year 1986.
(b) Effective 1/1/93, the Washington site no longer accepted waste from outside the Northwest and Rocky Mountain Compacts.
(c) Effective 7/1/2000, rates are based on whether a waste generator is or is not a member of the Atlantic Compact.
(d) Effective 7/1/2008, the South Carolina site no longer accepted waste from outside the Atlantic Compact.
(e) B_x values for the generic site are assumed to be the same as that provided for the Atlantic Compact, for lack of a better alternative at this time.
(f) Effective with NUREG-1307, Revision 8 (Ref. 3), an alternative disposal option was introduced in which the bulk of the LLW is assumed to be dispositioned by waste vendors and/or disposed of at a non-compact disposal facility.
(g) Effective with NUREG-1307, Revision 15, the nomenclature for the two disposal options, referred to as "Direct Disposal" and "Direct Disposal with Vendors" in previous revisions of NUREG-1307, is changed to "Compact-Affiliated Disposal Facility Only" and "Combination of Compact-Affiliated and Non-Compact Disposal Facilities" to better describe these options.

3. DEVELOPMENT OF COST ADJUSTMENT FORMULA

The evaluations presented in this chapter are based on information presented in NUREG/CR-0130, "Technology, Safety and Costs of Decommissioning a Reference Pressurized-Water Reactor Power Station–Technical Support for Decommissioning Matters Related to Preparation of the Final Decommissioning Rule," Addendum 4, and NUREG/CR-0672, "Technology, Safety and Costs of Decommissioning a Reference Boiling-Water Reactor Power Station–Technical Support for Decommissioning Matters Related to Preparation of the Final Decommissioning Rule," Addendum 3 (Refs. 1, 2), in which the estimated costs for immediate dismantlement of the reference PWR and the reference BWR are adjusted to January 1986 dollars. Decommissioning costs are divided into three general areas according to Title 10 of the *Code of Federal Regulations* (10 CFR) 50.75(c)(2) that tend to escalate similarly: (1) labor, materials, and services, (2) energy and waste transportation, and (3) radioactive waste burial/disposition. A relatively simple equation can be used to determine the minimum decommissioning fund requirement in year 2012 or previous year dollars. That equation is:

$$\text{Estimated cost (Year X)} = [1986 \ \$ \ \text{Cost}]*(A*L_x + B*E_x + C*B_x)$$

where:

Estimated cost (Year X) = estimated decommissioning costs in Year X dollars,

[1986 $ Cost] = estimated decommissioning costs in 1986 dollars,

A = fraction of the [1986 $ Cost] attributable to labor, materials, and services (0.65),

B = fraction of the [1986 $ Cost] attributable to energy and transportation (0.13),

C = fraction of the [1986 $ Cost] attributable to waste burial (0.22),

L_x = labor, materials, and services cost adjustment, January of 1986 to latest month of Year X for which data are available,

E_x = energy and waste transportation cost adjustment, January of 1986 to latest month of Year X for which data are available,

B_x = Low-level waste (LLW) burial/disposition cost adjustment, January of 1986 to the latest month of Year X for which data are available,

   = $(R_x + \Sigma S_x) / (R_{1986} + \Sigma S_{1986})$, where:

R_x = radioactive waste burial/disposition costs (excluding surcharges) in Year X dollars,

ΣS_x = summation of surcharges in Year X dollars,

R_{1986} = radioactive waste burial costs (excluding surcharges) in 1986 dollars, and

ΣS_{1986} = summation of surcharges in 1986 dollars.

Values for L_x and E_x for years subsequent to 1986 are to be based on the national producer price indexes, national consumer price indexes, and local conditions for a given site, as outlined in Sections 3.1 and 3.2. Thus, the licensee can evaluate these parameters appropriately for a particular site. The values to be used in determining B_x are taken from published cost schedules at the two compact-affiliated disposal facilities and the price quote for the non-compact disposal facility located in Clive, Utah.

Values of B_x for the year 2010, and earlier years, are provided to the licensees through this report for information purposes only.

The major elements of the three components of the decommissioning cost estimates for both the reference PWR and BWR are provided in Table 3-1. Considering the uncertainties and contingencies contained within these numbers, and considering that the values of the coefficients for the PWR and the BWR are so similar, the best estimates of their values are their averages for both the PWR and BWR estimates.

$$A_{ave} = 0.65 \qquad B_{ave} = 0.13 \qquad C_{ave} = 0.22$$

Table 3-1. Evaluation of the Coefficients A, B, and C in January 1986 Dollars

Cost Category	Reference PWR Values		Reference BWR Values	
	1986 $ (millions)	Coefficient	1986 $ (millions)	Coefficient
Labor	17.98[a]		35.12[b]	
Equipment	1.64[a]		4.03[b]	
Supplies	3.12[a]		3.71[b]	
Contractor	12.9[a]		21.1[b]	
Insurance	1.9[a]		1.9[b]	
Containers	10.9[d]		8.14[c]	
Added Staff	7.5[a]		4.4[b]	
Added Supplies	1.2[a]		0.2[b]	
Spec. Contractor	0.78[a]		0.71[b]	
Pre-engineering	7.4[a]		7.4[b]	
Post-TMI-backfits	0.9[a]		0.1[b]	
Surveillance	0.31[a]		--	
Fees	0.14[a]		0.14[b]	
Subtotal	**66.67**	**A = 0.64**	**86.95**	**A = 0.66**
Energy	8.31[a]		8.84[b]	
Transportation	6.08[d]		7.54[c]	
Subtotal	**14.39**	**B = 0.14**	**16.38**	**B = 0.12**
Burial	22.48[d]	C = 0.22	29.98[c]	C = 0.22
Total	**103.54**		**133.31**	

Note: All costs include a 25% contingency factor.
(a) Based on Table 3.1, NUREG/CR-0130, Addendum 4.
(b) Based on Table 3.1, NUREG/CR-0672, Addendum 3.
(c) Based on Table 5.2, NUREG/CR-0672, Addendum 3.
(d) Based on Table 6.2, NUREG/CR-0130, Addendum 4.

3.1 Labor Adjustment Factors

Current employment cost indexes for labor (column 3, Table 3-2, below) can be obtained from the "Employment Cost Indexes," published by the U.S. Department of Labor, Bureau of Labor Statistics (BLS) (Ref. 4). Specifically, the appropriate regional data from Table 6 of Reference 4 entitled "Employment Cost Index for total compensation, for private industry workers, by bargaining status, census region and division, and metropolitan area status" should be used. These indexes also may be obtained from BLS databases available on the Internet (see APPENDIX C for instructions).

To calculate the current labor adjustment factor (L_x) for a particular region, two numbers are needed: a base labor adjustment factor and the current Employment Cost Index (ECI). The base labor adjustment factors are shown in column 2 of Table 3-2, and the current ECIs are shown in column 3. The base labor adjustment factor is the value of L_x at the time the ECI was most recently re-indexed. (This latest re-indexing occurred in December 2005, at which time the index was reset to 100.) As such, current values of L_x (column 4) are obtained from the simple proportion:

$$L_x/ECI = \text{Base } L_x/100$$

For example, for the Northeast region,

$$L_x/116.5 = 2.16/100$$

or

$$L_x = 2.16*116.5/100 = 2.52$$

Table 3-2. Regional Factors for Labor Cost Adjustment

Region	Base L_x (Dec 2005)	Qtr 1 2012 ECI (Dec 2005 = 100)	L_x (Qtr 1 2012)
Northeast	2.16	116.5	2.52
South	1.98	116.0	2.30
Midwest	2.08	114.7	2.39
West	2.06	115.7	2.38

3.2 Energy Adjustment Factors

The adjustment factor for energy, E_x, is a weighted average of two components, namely, industrial electric power, P_x, and light fuel oil, F_x. For the reference PWR, E_x is given by:

$$E_x \text{ (PWR)} = 0.58P_x + 0.42F_x$$

and for the reference BWR it is:

$$E_x \text{ (BWR)} = 0.54P_x + 0.46F_x$$

These equations are derived from Table 6-3 of Reference 1 and Table 5-3 of Reference 2. The current values of P_x and F_x are calculated from the Producer Price Indexes (PPI), available in the "PPI Detailed Report," published by the U.S. Department of Labor, BLS (Ref. 5). These indexes also can be obtained from BLS databases available on the Internet (see APPENDIX C for instructions). The indexes used to calculate P_x should be taken from data for industrial electric power (PPI Commodity Code 0543), and the indexes used to calculate F_x should be taken from data for light fuel oils (PPI Commodity Code 0573). No regional BLS data for these PPI commodity codes are currently available.

P_x and F_x are the values of current producer price indexes (PPI Codes 0543 and 0573, respectively) divided by the corresponding indexes for January 1986. All PPI values are based on a value of 100 for the year 1982 (base 1982 = 100). Thus, the values of P_x and F_x for March 2012 (latest data available) are:

P_x = 199.8 (March 2012 value of code 0543) ÷
      114.2 (January 1986 value of code 0543) = 1.750

F_x = 329.8 (March 2012 value of code 0573) ÷
      82.0 (January 1986 value of code 0573) = 4.022

The value of E_x for the reference PWR is therefore

E_x (PWR) = [(0.58 x 1.750) + (0.42 x 4.022)] = 2.704.

This value of E_x = 2.704 should then be used in the equation to adjust the energy cost (to March 2012 dollars) for decommissioning a PWR.

For the reference BWR,

E_x (BWR) = [(0.54 x 1.750) + (0.46 x 4.022)] = 2.795.

3.3 Waste Burial Adjustment Factors

The adjustment factor for waste burial/disposition, B_x, can be taken directly from data on the appropriate LLW burial location as given in Table 2-1 of this report. For example, B_x = 30.581 (in 2012 dollars) for a PWR in the Atlantic Compact disposing all decommissioning waste at the compact-affiliated disposal facility.

3.4 Sample Calculations of Estimated Reactor Decommissioning Costs

Four sample calculations are provided in this section to demonstrate the use of the decommissioning cost equation developed above using the appropriate adjustment terms of L_x for labor, material, and services; E_x for energy and waste transportation; and B_x for radioactive waste burial/disposition. The coefficients A, B, and C (0.65, 0.13, and 0.22, labor, energy, and burial fractions, respectively) used in the examples are developed in Table 3-1. Waste generators with no disposal site available for LLW should use the B_x values for the generic LLW disposal site. Sample decommissioning costs for other years are provided in APPENDIX D.

Example 1 (Compact-Affiliated Disposal Facility Only)

Scenario Description
  Reactor Type: PWR
  Thermal Power Rating: 3,400 megawatt thermal (MWth)
  Location of Plant: Northwest Compact
  LLW Disposition Preference: Compact-Affiliated Disposal Facility Only
  LLW Burial Location: Washington

Base Cost (1986 Dollars) = $105 million [from 10 CFR 50.75(c)(1)]

L_x =   2.38 [from Table 3-2]

E_x =   2.704 [from Section 3.2]

B_x =   7.335 [from Table 2-1]

Decommissioning Cost (2012 dollars)
  = ($105 million)*[(0.65)*(2.38)+(0.13)*(2.704)+(0.22)*(7.335)]
  = $369 million

Example 2 (Compact-Affiliated Disposal Facility Only)

Scenario Description
  Reactor Type: PWR
  Thermal Power Rating: 3,400 MWth
  Location of Plant: Atlantic Compact
  LLW Disposition Preference: Compact-Affiliated Disposal Facility Only
  LLW Burial Location: South Carolina (Atlantic Compact)

Base Cost (1986 Dollars) = $105 million [from 10 CFR 50.75(c)(1)]

L_x =   2.52 [from Table 3-2]

E_x =   2.704 [from Section 3.2]

B_x =   30.581 [from Table 2-1]

Decommissioning Cost (2012 dollars)
  = ($105 million)*[(0.65)*(2.52)+(0.13)*(2.704)+(0.22)*(30.581)]
  = $915 million

Example 3 (Combination of Compact-Affiliated and Non-Compact Disposal Facilities)

Scenario Description
    Reactor Type: PWR
    Thermal Power Rating: 3,400 MWth
    Location of Plant: Atlantic Compact
    LLW Disposition Preference: Combination of Compact-Affiliated and Non-Compact Disposal Facilities
    LLW Burial Location: South Carolina (Atlantic Compact)

Base Cost (1986 Dollars) = $105 million [from 10 CFR 50.75(c)(1)]

L_x =   2.52 [from Table 3-2]

E_x =   2.704 [from Section 3.2]

B_x =   13.885 [from Table 2-1]

Decommissioning Cost (2012 dollars)
  = ($105 million)*[(0.65)*(2.52)+(0.13)*(2.704)+(0.22)*(13.885)]
  = $530 million

Example 4 (Combination of Compact-Affiliated and Non-Compact Disposal Facilities)

Scenario Description
    Reactor Type: BWR
    Thermal Power Rating: 3,400 MWth
    Location of Plant: Midwest Compact
    LLW Disposition Preference: Combination of Compact-Affiliated and Non-Compact Disposal Facilities
    LLW Burial Location: Unknown (Generic LLW Disposal Site)

Base Cost (1986 Dollars) = $135 million [from 10 CFR 50.75(c)(1)]

L_x =   2.39 [from Table 3-2]

E_x =   2.795 [from Section 3.2]

B_x =   14.160 [from Table 2-1]

Decommissioning Cost (2012 dollars)
  = ($135 million)*[(0.65)*(2.39)+(0.13)*(2.795)+(0.22)*(14.160)]
  = $679 million

4. **REFERENCES**

1. Konzek G.J. and R.I. Smith, "Technology, Safety, and Costs of Decommissioning a Reference Pressurized-Water Reactor Power Station–Technical Support for Decommissioning Matters Related to Preparation of the Final Decommissioning Rule," (Report prepared by Pacific Northwest Laboratory, Richland, Washington), NUREG/CR-0130, Addendum 4, U.S. Nuclear Regulatory Commission, July 1988.

2. Konzek G.J. and R.I. Smith, "Technology, Safety and Costs of Decommissioning a Reference Boiling-Water Reactor Power Station–Technical Support for Decommissioning Matters Related to Preparation of the Final Decommissioning Rule," (Report prepared by Pacific Northwest Laboratory, Richland, Washington), NUREG/CR-0672, Addendum 3, U.S. Nuclear Regulatory Commission, July 1988.

3. U.S. Nuclear Regulatory Commission, Office of Nuclear Regulatory Research, "Report on Waste Burial Charges–Changes in Decommissioning Waste Disposal Costs at Low-Level Waste Burial Facilities," NUREG-1307, Revision 8, December 1998.

4. U.S. Department of Labor, Bureau of Labor Statistics, *Employment Cost Indexes*, updated annually (approximately) through various bulletins.

5. U.S. Department of Labor, Bureau of Labor Statistics, *PPI Detailed Report*, updated monthly.

APPENDIX A.

LOW-LEVEL WASTE BURIAL/DISPOSITION
PRICES FOR THE CURRENT YEAR

APPENDIX A.

LOW-LEVEL WASTE BURIAL/DISPOSITION PRICES FOR THE CURRENT YEAR

This appendix contains the price schedules for burial/disposition of LLW at the Washington and South Carolina sites for the year 2012. Also provided is a price quote for the non-compact disposal facility located in Clive, Utah. These schedules are used to calculate the burial/disposition costs discussed in APPENDIX B.

A.1 Washington LLW Disposal Site

Beginning in 1993, the Northwest Compact imposed on eligible (Northwest or Rocky Mountain Compact) waste generators an annual permit fee based on the volume of waste to be shipped to the Washington site for disposal. For 2012, the permit fees range from $424 to $42,400. Hospitals, universities, research centers, and industries pay the lower fees; NPPs pay the highest fee of $42,400. Permit fees for NPPs are included in this analysis for the years 1993 and later.

Beginning in 1994, the rate schedule for handling and disposing of heavy objects (greater than 5,000 pounds) at the Washington site was revised to recover additional crane rental costs from the waste generator. In 1996, the heavy object limit was raised to 17,500 pounds. A series of shipments of heavy objects for disposal was assumed that would minimize the crane surcharge and result in a one-time only heavy object charge.

Effective January 1, 1996, the operator of the Washington site implemented a restructured rate schedule based on waste volume, number of shipments, number of containers, and dose rate at the container surface. Each waste generator is also assessed an annual site availability charge based on cumulative volume and dose rate at the surface of all containers disposed. The site availability charge appears near the bottom of Table B-1 through Table B-12.

In 2000, charges for all ranges of container surface dose rates were reduced by a factor of eight compared to 1998. This significantly reduced burial costs at the Washington LLW disposal site. However, effective May 1, 2002, these surface dose rate charges had increased by more than a factor of eight (to about what they were in 1998). In addition, volume, shipment, and container charges had increased by 6.5 percent, 42.2 percent, and 42.2 percent, respectively. Thus, burial charges for 2002 were significantly higher than the charges for 2000, but they are roughly comparable to what they were in 1998.

The 2008 rate schedule reflects increases in volume (14 percent), shipment (22 percent), and container (17 percent) charges compared to 2006. In addition, dose rate charges per container increased by a factor of 2.8. As a result of these changes, the cost to disposition a PWR increased moderately, 21 percent. However, the cost to disposition a BWR, with its larger volume of high dose rate material, almost doubled.

In 2010, two algorithm changes were implemented to project more accurately charges for waste generated from the decommissioning of an NPP. The first was a discount to the volume disposal rate of 20 percent for LLW generated from the decommissioning of NPPs. The second was to cap the container dose rate charge. According to the settlement agreement between

U.S. Ecology Washington, Inc., the operator of the Washington disposal facility, and the State of Washington, only 14.2 percent of the Washington site's revenue requirement (which changes annually) may be recouped from container dose rate charges.

Compared with the 2010 rate schedule used in Revision 14 of this report, the 2012 schedule reflects decreases in volume (9 percent) and container (16 percent) charges and an increase in shipment (3 percent) charges. In addition, dose rate charges per container increased by a factor of 5.4. As a result of these rate changes, the cost to disposition the LLW from a PWR decreased by 9 percent and the cost to disposition the LLW from a BWR decreased by 10 percent. The rate schedule for the Washington LLW disposal site, effective May 1, 2012, is presented in Exhibit A-1.

A.2   South Carolina LLW Disposal Site

Access to the South Carolina site by waste generators outside the Southeast Compact ended June 30, 1994, with site closure scheduled for December 31, 1995. However, effective July 1, 1995, the scheduled closure was canceled and access to the Barnwell facility was extended to all States except North Carolina. In June 2000, prohibition on waste from North Carolina was lifted.

Effective November 1, 1996, the operator of the South Carolina disposal site implemented a restructured waste disposal rate schedule. The restructured pricing is based on weight, dose rate, and curies with a cost incentive toward higher density packaging. All business after November 1, 1996, is through customer-specific contracts.

From July 1, 1998, through June 30, 1999, the operator of the South Carolina disposal site imposed a site access fee on users, which varied according to their level of use. Access fees for large users (e.g., utilities with nuclear plants) averaged about $205,000 for the year.

In the years between 2001 and 2008, the maximum allowable volume of LLW disposed of at the South Carolina LLW disposal site from all sources was governed by a schedule contained in the Atlantic Interstate Low-Level Radioactive Waste Compact Implementation Act, which was enacted into law July 1, 2000. This schedule is shown in Table A-1.

Table A-1. Schedule of Maximum Allowable LLW Disposal at the South Carolina Disposal Facility[a]

Fiscal Year	Maximum Allowable LLW Volume from All Sources (ft^3)
2001	160,000
2002	80,000
2003	70,000
2004	60,000
2005	50,000
2006	45,000
2007	40,000
2008	35,000

(a) Reference: Code of Laws of South Carolina, 1976, Section 1, Title 48, Chapter 46.

Effective July 1, 2008, out-of-compact waste was prohibited from disposal at the South Carolina disposal site.

Weight charges, curie surcharges, and irradiated hardware charges all increased approximately 12 percent from the 2010 Atlantic Compact rates, while dose rate and administrative surcharge remained constant. As a result of these changes, the cost to disposition the LLW from both a BWR and a PWR increased approximately 12 percent. The rate schedule for the South Carolina LLW disposal site, effective July 1, 2012, is presented in Exhibit A-2.

A.3   Alternative LLW Disposal Options

In the 1990s rapidly increasing fees for disposal of low-level radioactive waste spawned the creation of a niche market for firms specializing in the management and disposal of LLW. Increasingly, NPP licensees began to outsource LLW management functions to waste vendors for a negotiated fee (usually $/pound of LLW processed) and disposing of Class A LLW at the non-compact disposal facility in Clive, Utah. Waste vendors could manage waste from generation to disposal (including packaging, transportation, and volume reduction) or any subset of these functions that the licensee desired.

The vendor determined the most efficient disposition process for each waste stream, which may have included sorting into clean and contaminated streams, recycling where possible, volume reduction through the many techniques currently commercially available, and disposal of the residual LLW at the most cost-effective disposal site, including the non-compact disposal facility located in Clive, Utah. The vendor's profit was the difference between the price negotiated with the licensee and the total cost for waste minimization, recycling, volume reduction, packaging, transportation, and disposal. The more effective the vendor was at minimization, recycling, volume reduction, and obtaining volume discounts for packaging, transportation, and disposal, the greater its profit would be.

The decommissioning analyses reported in NUREG/CR-0130 and NUREG/CR-0672 did not consider the possible use of waste vendors or non-compact Class A LLW disposal facilities, given that these market niches essentially did not exist at the time. Beginning with Revision 8, NUREG-1307 included an alternative that provided for contracting with waste vendors to manage the disposition the bulk of LLW generated during decommissioning. This new alternative did not modify or alter in any way the bases for the decommissioning fund requirement specified in 10 CFR 50.75, "Reporting and Recordkeeping for Decommissioning Planning." It merely provided an alternative burial cost adjustment factor (B_x) that reflected the option of disposing of LLW using a combination of waste vendors, non-compact disposal facilities, and compact-affiliated disposal facilities.

In support of the analysis performed for NUREG-1307, Revision 8 (Ref. 3), several waste vendors were surveyed to develop a representative cost for waste vendor services. Each vendor was asked to provide a generic price quote for processing two waste streams: activated and contaminated concrete and contaminated metal. Vendors were asked to provide these quotes as a price per pound of waste, or as a range of prices per pound, based on the waste concrete and metal inventories in NUREG/CR-0130 and NUREG/CR-0672. The price quotes were to encompass complete disposition of these waste streams (from generation to disposal) and to be developed assuming the vendor had a contract with a licensee engaged in a large decommissioning project.

Based on the results of the survey, NUREG-1307, Revision 8, introduced an alternative burial cost adjustment factor (B_x) that assumed the use of waste vendor services and disposal of Class A LLW at the non-compact disposal facility located in Clive, Utah as an alternative to disposal of all decommissioning LLW at a compact-affiliated disposal facility. The option was introduced to provide potential savings from the use of waste vendors. For a PWR under this option, 93% of the waste was assumed to be dispositioned by waste vendors and the remaining 7% was assumed to be disposed of at a compact-affiliated disposal facility. For a BWR under this option, 95% of the waste was assumed to be dispositioned by waste vendors and the remaining 5% was assumed to be disposed of at a compact-affiliated disposal facility. These proportions were determined from a component-by-component analysis of the reference BWR and PWR. The portions of waste assumed to be dispositioned by waste vendors were priced at the rates obtained from the vendor surveys, and the portions of waste assumed to be disposed of at compact-affiliated disposal facilities were priced at rates obtained for those facilities.

In support of Revision 15 of NUREG-1307, a similar survey was conducted. In response to this survey, a price quote to disposition the components of the reference PWR and BWR at the Utah disposal facility was obtained. Unit costs, exclusive of taxes, were provided for several different categories of components, which are provided in Table A-2. The updated rates represent an average increase of 15.5 percent compared to the 2010 rates. These rates assume no volume discounts, which can be substantial. The development of the B_x factor for the "Combination of Compact-Affiliated and Non-Compact Disposal Facilities" option was based on these rates and an assumed a 10 percent tax.

Table A-2. Price Quotes for Disposition of Class A LLW at the Non-Compact Disposal Facility Located in Clive, Utah

Component Class	Cost	Per Unit
Large Components	$350	ft^3
Debris	$145	ft^3
Oversize Debris	$165	ft^3
Resins/Filters	$460	ft^3
Combustibles	$575	ft^3
Evaporator Bottoms	$14	gal

Exhibit A-1

U.S. ECOLOGY WASHINGTON, INC.
RICHLAND, WASHINGTON FACILITY
RADIOACTIVE WASTE DISPOSAL

SCHEDULE OF CHARGES
EFFECTIVE MAY 1, 2012
SCHEDULE A, 16th REVISION

Note: Rates in this Schedule A are subject to adjustment in accordance with the rate adjustment mechanism adopted in the Washington Utilities and Transportation Commission's Sixth Supplemental Order in Docket No. UR·950619 as extended by Commission Order in Docket Nos. UR·010623 and UR-010706, and TL-070848.

A.   SITE AVAILABILITY CHARGE

1.   Rates

Block Criteria	Annual Charge per Generator
0 No site use at all	$251
1 Greater than zero but less than or equal to 10 ft^3 and 50 mR/h	481
2 Greater than 10 ft^3 or 50 mR/h* but less than or equal to 20 ft^3 and 100 mR/h*	924
3 Greater than 20 ft^3 or 100 mR/h* but less than or equal to 40 ft^3 and 200 mR/h*	1,773
4 Greater than 40 ft^3 or 200 mR/h* but less than or equal to 80 ft^3 and 400 mR/h*	3,404
5 Greater than 80 ft^3 or 400 mR/h* but less than or equal to 160 ft^3 and 800 mR/h*	6,538
6 Greater than 160 ft^3 or 800 mR/h* but less than or equal to 320 ft^3 and 1,600 mR/h*	12,539
7 Greater than 320 ft^3 or 1,600 mR/h* but less than or equal to 640 ft^3 and 3,200 mR/h*	24,077
8 Greater than 640 ft^3 or 3,200 mR/h* but less than or equal to 1,280 ft^3 and 6,400 mR/h*	46,221
9 Greater than 1,280 ft^3 or 6,400 mR/h* but less than or equal to 2,560 ft^3 and 12,800 mR/h*	88,743
10 Greater than 2,560 ft^3 or 12,800 mR/h* but less than or equal to 5,120 ft^3 and 25,600 mR/h*	133,026
11 Greater than 5,120 ft^3 or 25,600 mR/h*	133,026

* For purposes of determining the site availability charge, mR/hour is calculated by summing the mR per hour at container surface of all containers received during the year.

2.   Exemptions

a.   As to waste which is generated by educational research institutions for research, medical or educational purposes, such institutions shall be placed in a rate block for the site availability charge which is one (1) lower than what would otherwise apply through application of the block criteria shown above. *"Educational research Institution" means a state or independent, not-for-profit, post-secondary educational institution.*

b.   As to waste which arises as residual or secondary waste from brokers' provision of compaction or processing services for others, if application of the block criteria shown above would place a broker in a rate block for the site availability charge which is greater than Block No. 7, such broker shall be placed in the rate block which is the greater of (i) Block No. 7, or (ii) the block which is two (2) lower than what would otherwise apply through application of the block criteria shown above. "Brokers" are those customers holding the "broker" classification of site use permits issued by the Department of Ecology.

3.   Payment Arrangements

    a.   Initial Determination

    Initial determination as to the applicable rate block for each customer shall be based on projections provided by customers prior to the beginning of each calendar year. For customers who do not intend to ship waste to the facility during the calendar year (those assigned to block No. 0) and for those customers who are initially determined to fall into block Nos. 1–2, the entire site availability charge for the year will be due and payable as of January 1. For those customers who are initially determined to fall into block Nos. 3–8, the entire site availability charge will be due and payable as of January 1, although those customers may make special arrangements with the Company to pay the charge in equal installments at the beginning of each calendar quarter. For those generators who are initially determined to fall in block Nos. 9-11, 1/12 of the site availability charge will be due and payable as of the beginning of each calendar month. These customers may pay in advance if they wish.

    b.   Reconciliation

    The site availability charge is assessed on the basis of actual volume and dose rate of waste delivered during the calendar year.

    Assessment of additional amounts, or refunds of overpaid amounts, will be made as appropriate to reconcile the initial determination regarding applicable rate block with the actual volume and dose rates during the calendar year.

Exhibit A-1

SCHEDULE A (Continued)

B.   DISPOSAL RATES

1.   Volume: $115.50 per cubic foot

2.   Shipment: $13,750 per manifested shipment

3.   Container: $7,560 per container on each manifest.

4.   Exposure:

Block No.	Dose Rate at Container Surface	Charge per Container
1	Less than or equal to 200 mR/h	$92
2	Greater than 200 mR/h but less than or equal to 1,000 mR/h	6,540
3	Greater than 1,000 mR/h but less than or equal to 10,000 mR/h	26,200
4	Greater than 10,000 mR/h but less than or equal to 100,000 mR/h	39,300
5	Greater than 100,000 mR/h	661,500

EXTRAORDINARY VOLUMES
Waste shipments qualifying as an "extraordinary volume" under RCW 81.108.020(3) are charged a rate equal to 51.5 percent of the volume disposal rate.

NUCLEAR DECOMMISSIONING WASTE
The volume disposal rate applicable to waste from the decommissioning of nuclear generating units shall be 80 percent of those set forth above; provided, however, that such waste must satisfy the quantity requirements for "extraordinary volume" under RCW 81.108.020(3).

SCHEDULE B
Surcharges and Other Special Charges
Eighth Revision

ENGINEERED CONCRETE BARRIERS
        72" x 8' barrier $11,432.00 each
        84" x 8' barrier $13,987.00 each

SURCHARGE FOR HEAVY OBJECTS
The Company shall collect its actual labor and equipment costs incurred, plus a margin thereon of 25 percent, in handling and disposing of objects or packages weighing more than seventeen thousand five hundred (17,500 pounds.

SCHEDULE C
Tax and Fee Rider
Original Tariff

The rates and charges set forth in Schedules A and B shall be increased by the amount of any fee, surcharge, or tax assessed on a volume or gross revenue basis against or collected by U.S. Ecology Washington, Inc. as listed below:

Perpetual Care and Maintenance Fees	$1.75 per cubic foot
Business & Occupation Tax	3.3 percent of rates and charges
Site Surveillance Fee	$9.00 per cubic foot
Surcharge (RCW 43.200.233)	$6.50 per cubic foot
Commission Regulatory Fee	1.0 percent of rates and charges

Exhibit A-2

Pursuant to 48-46-40(A)(2), S.C.C.

Uniform Schedule of Maximum Disposal Rates
for Atlantic Compact Regional Waste

EFFECTIVE JULY 1, 2012

The Uniform Schedule of Maximum Disposal Rates for Atlantic Compact Regional Waste is a permanent ceiling on disposal rates applicable to Atlantic Compact waste that is adjusted each year in accordance with the Producer Price Index. South Carolina may charge Atlantic Compact generators less than the Uniform Maximum schedule, but cannot charge regional generators more than this rate.

THE MINIMUM CHARGE PER SHIPMENT, EXCLUDING SURCHARGES AND SPECIFIC OTHER CHARGES, IS $1,000.00

1. **WEIGHT CHARGES (not including surcharges)**

   A. **Base weight charge**

Density Range	Weight Rate
i) Equal to or greater than 120 lbs./ft^3	$7.516 per pound
ii) Equal to or greater than 75 lbs./ft^3 and less than 120 lbs./ft^3	$8.268 per pound
iii) Equal to or greater than 60 lbs./ft^3 and less than 75 lbs./ft^3	$10.148 per pound
iv) Equal to or greater than 45 lbs./ft^3 and less than 60 lbs./ft^3	$13.155 per pound
iv) Less than 45 lbs./ft^3	$13.155 per pound multiplied by: (45 ÷ pounds per cubic foot of the package)

   B. **Dose multiplier on base weight charge**

Container Dose Level	Multiplier on Weight Rate, above
0 mR/hr - 200 mR/hr	1.00
>200 mR/hr - 1 R/hr	1.08
>1R/hr - 2R/hr	1.12
>2R/hr - 3R/hr	1.17
>3R/hr - 4R/hr	1.22
>4R/hr - 5R/hr	1.27
>5R/hr - 10R/hr	1.32
>10R/hr - 25R/hr	1.37
>25R/hr - 50R/hr	1.42
>50R/hr	1.48

   C. **Biological Waste:** Add $1.523 per pound to rate calculated above

2.   SURCHARGES

A.   Millicurie surcharge $.563 per millicurie*

*In lieu of above, generator may opt for an alternative millicurie charge of $1.125 per millicurie applicable only to millicuries with greater than a 5-year half-life. Such election must be provided in writing to the disposal site operator prior to July 1, 2012.

MAXIMUM MILLICURIE CHARGE IS $225,083 PER SHIPMENT (400,000 MCI).

B.   Irradiated hardware charges $85,422 per shipment
     (See Note B under Miscellaneous)

C.   Special nuclear material surcharge $17.082 per gram

D.   Atlantic Compact Commission administrative surcharge $6 per cubic foot

(Subject to change during year)

Exhibit A-2

NOTES

A. Surcharges for the Barnwell Extended Care Fund and the Decommissioning Trust Fund are included in the rates.

B. Irradiated hardware: As a general rule, billing as irradiated hardware pertains to shipments of exceptionally high activity that require clearing of the site and special offloading into a slit trench. These generally include CNS3-55, TN-RAM, and other horizontally off-loaded cask shipments. In addition to items of irradiated hardware, shipments considered irradiated hardware, for purposes of disposal, have included certain sealed sources and materials with exceptionally high levels of radioactivity.

C. Large components (e.g., steam generators, reactor pressure vessels, coolant pumps).

Disposal fees for large components (e.g., steam generators, reactor pressure vessels, reactor coolant pumps, or items that will not fit into standard sized disposal vaults) are based on the generally applicable rates, in their entirety, except that the weight and volume used to determine density and weight related charges is calculated as follows:

1. For packages where the large component shell qualifies as the disposal vault per Department of Health and Environmental Control (DHEC) regulations, weight and volume calculations are based on all subcomponents and material contained within the inside surface of the large component shell, including all internals and any stabilization media injected by the shipper, but excluding the shell itself and all incidental external attachments required for shipping and handling; and

2. For packages with a separate shipping container that qualifies as the disposal vault per DHEC regulations, weight and volume calculations are based on the large component, all subcomponents, and material contained within the inside surface of the shipping container, including any stabilization media injected by the shipper (including that between the large component and the shipping container), but excluding the shipping container itself and all incidental external attachments required for shipping and handling.

D. Co-mingled shipments from brokers and processors: For containers that include waste from different generators (DHEC permittees), the weight and density of the waste from each generator will be assessed separately for purposes of the weight charge in I.A. The dose of the container as a whole will be used to assess the dose multiplier in I.B. The millicurie charge 2.A. above applies individually to each portion of waste in the shipment from each generator. The disposal site operator will provide guidelines for application of this method.

E. Transport vehicles with additional shielding features may be subject to an additional handling fee, which will be provided upon request.

F. In certain circumstances, the disposal site operator may assess additional charges for necessary services that are not part of and are in addition to disposal rates established by the State of South Carolina. These include decontamination services and special services as described in the Barnwell Site Disposal Criteria.

G. The disposal site operator has established the following policies and procedures, which are provided herein for informational purposes:

i. Terms of payment are net 30 days upon presentation of invoices. A per-month service charge of one and one-half percent (1½ percent) shall be levied on accounts not paid within 30 days.

ii. Company purchase orders or a written letter of authorization and substance acceptable to Chem-Nuclear Systems, L.L.C. (CNS) shall be received before receipt of radioactive waste material at the Barnwell Site and shall refer to CNS Radioactive Material License, the Barnwell Site Disposal Criteria, and subsequent changes thereto.

iii. All shipments shall receive a CNS shipment identification number and conform to the prior notification plan.

APPENDIX B.

CALCULATION OF LOW-LEVEL WASTE BURIAL/DISPOSITION COST ESTIMATION FACTORS

APPENDIX B.

CALCULATION OF LOW-LEVEL WASTE BURIAL/DISPOSITION COST ESTIMATION FACTORS

The calculations necessary to determine the costs for burial/disposition of radioactive wastes resulting from decommissioning the reference PWR and the reference BWR are performed using spreadsheet models. The spreadsheets evaluate the burial/disposition costs for each of the items originally budgeted in the PWR and BWR decommissioning studies and in Addendums 4 and 3 (Refs. 1, 2), respectively, to those reports. The costs are based on the published price schedules from the compact-affiliated disposal facilities and a price quote from the non-compact disposal facility located in Clive, Utah.

The B_x values reported in this document reflect the updated rate schedules and price quote. All the calculations are based on the same inventory of radioactive wastes as was postulated in the 1986 and 1978–1980 analyses. Starting in 1988, the inventories also included post-Three Mile Island (TMI)-2 contributions from the reference PWR and the reference BWR (Refs. 1, 2).

B.1 Washington LLW Disposal Site

The LLW disposal site located in Washington was used to develop the original decommissioning cost estimates for the reference PWR and the reference BWR. These estimates are the basis for the minimum decommissioning fund requirement specified in Title 10 of the *Code of Federal Regulations* (10 CFR) 50.75(c), which is in 1986 dollars. Thus, B_x = 1.0/1.0 (for PWR/BWR) for 1986.

For the year 2012, B_x = 7.335/6.704 for a PWR and BWR, respectively, at the Washington disposal facility. These B_x values reflect the adjustment in waste burial costs at the Washington LLW disposal site since 1986. B_x values for several previous revisions of NUREG-1307 are summarized in Table 2-1.

Waste burial costs for the year 2012 were developed using the rate schedule provided in Exhibit A.1. The spreadsheet calculations for the current year, which are too voluminous to present here, are summarized in Table B-1 and Table B-2. Table B-3 through Table B-12 provide summaries of the waste burial costs at the Washington LLW disposal site for 2010, 2008, 2006, 2004, and 2002, respectively. These estimates originally were reported in previous revisions of NUREG-1307.

B.2 South Carolina LLW Disposal Site

For the year 2012, B_x = 30.581/27.295 for a PWR and BWR, respectively, at the South Carolina disposal facility. These B_x values reflect the adjustment in waste burial costs at the South Carolina LLW disposal site normalized to the 1986 Washington LLW disposal site burial costs. B_x values for several previous revisions of NUREG-1307 are summarized in Table 2-1.

Waste burial costs for the year 2012 were developed using the rate schedules provided in Exhibit A.2. The spreadsheet calculations for the current year, which are too voluminous to present here, are summarized in Table B-13 and Table B-14. Table B-15 through Table B-30 provide summaries of the waste burial costs at the South Carolina LLW disposal site for 2010, 2008, 2006, 2004, and 2002, respectively. These estimates originally were reported in previous revisions of NUREG-1307.

B.3   Combination of Non-Compact and Compact-Affiliated Disposal Facilities

For the year 2012, B_x = 7.375/6.076 for disposal of most Class A LLW at the Utah non-compact site and the remaining LLW at the Washington LLW disposal site. B_x = 13.885/14.160 for disposal of most Class A LLW at the Utah non-compact site and the South Carolina compact-affiliated disposal site. B_x values are summarized in Table 2-1.

Waste burial costs for the year 2012 were developed using both the rate schedules for the compact-affiliated disposal facilities provided in Exhibits A-1 and A-2, and the rate quote for the non-compact disposal facility provided in Table A-2. The spreadsheet calculations for the current year, which are too voluminous to present here, are summarized in Table B-31 through Table B-34. Table B-35 through Table B-60 provide summaries of the waste burial/disposition costs for 2010, 2008, 2006, 2004, and 2002, respectively. These estimates were originally reported in previous revisions of NUREG-1307.

B.4   Other

As other low-level radioactive waste burial sites come into service in the interstate compacts, values for B_x will be calculated using the price schedules for each of those sites and will be incorporated into subsequent issues of this report. Those materials whose activity concentrations exceed the limits for Class C LLW are identified by footnote as greater-than-Class C (GTCC) material. Because the analyses in this report postulate placing this material in a LLW disposal facility, the disposal costs for this material may be significantly overestimated compared with high-density packaging and geologic repository disposal. It may also be feasible to store GTCC waste in independent spent fuel storage installations (ISFSIs) or other interim storage facilities, as permitted by 10 CFR Part 72, "Licensing Requirements for the Independent Storage of Spent Nuclear Fuel, High-Level Radioactive Waste, and Reactor-Related Greater Than Class C Waste."

Table B-1. PWR Burial Costs at the Washington Site (2012 dollars)

REFERENCE PWR COMPONENT	VOLUME CHARGE	SHIPMENT CHARGE	CONTAINER CHARGE	CONTAINER DOSE RATE CHARGE	DISPOSAL COST
VESSEL WALL	351,120	522,500	287,280	777,649	1,938,549
VESSEL HEAD & BOTTOM	369,600	550,000	302,400	0	1,222,000
UPPER CORE SUPPORT ASSM	36,960	55,000	30,240	0	122,200
UPPER SUPPORT COLUMN	36,960	55,000	30,240	0	122,200
UPPER CORE BARREL	18,480	27,500	15,120	0	61,100
UPPER CORE GRID PLATE	46,200	68,750	37,800	0	152,750
GUIDE TUBES	55,440	82,500	45,360	0	183,300
LOWER CORE BARREL[a]	295,680	440,000	241,920	0	977,600
THERMAL SHIELDS[a]	55,440	82,500	45,360	0	183,300
CORE SHROUD[a]	36,960	55,000	30,240	0	122,200
LOWER GRID PLATE[a]	46,200	68,750	37,800	0	152,750
LOWER SUPPORT COLUMN	9,240	13,750	7,560	0	30,550
LOWER CORE FORGING	101,640	151,250	83,160	0	336,050
MISC INTERNALS	73,920	110,000	60,480	0	244,400
BIO SHIELD CONCRETE	2,306,304	673,750	1,474,200	0	4,454,254
REACTOR CAVITY LINER	47,309	13,750	30,240	0	91,299
REACTOR COOLANT PUMPS	388,080	165,000	90,720	0	643,800
PRESSURIZER	332,640	110,000	60,480	0	503,120
R.Hx,EHx,SUMP PUMP, CAVITY PUMP	36,960	13,750	22,680	0	73,390
PRESSURIZER RELIEF TANK	110,880	27,500	15,120	0	153,500
SAFETY INJECTION ACCUM TANKS	369,600	110,000	60,480	0	540,080
STEAM GENERATORS	1,973,849	440,000	241,920	0	2,655,769
REACTOR COOLANT PIPING	304,920	96,250	52,920	0	454,090
REMAINING CONTAM. MATLS	4,860,979	1,388,750	3,107,160	0	9,356,889
CONTAMINATED MATRL OTHR BLD	44,085,056	10,945,000	28,040,040	0	83,070,096
FILTER CARTRIDGES	29,106	82,500	45,360	0	156,966
SPENT RESINS	184,800	275,000	151,200	0	611,000
COMBUSTIBLE WASTES	935,550	825,000	453,600	0	2,214,150
EVAPORATOR BOTTOMS	868,560	1,292,500	710,640	0	2,871,700
POST-TMI-2 ADDITIONS	1,438,021	0	0	0	1,438,021
HEAVY OBJECT SURCHARGE					171,988
SITE AVAILABILITY CHARGES					399,078
SUBTOTAL PWR COSTS	**59,806,454**	**18,741,250**	**35,811,720**	**777,649**	**115,708,139**
TAXES & FEES (% OF CHARGES)					4,975,450
TAXES & FEES ($/UNIT VOL.)					11,130,666
ANNUAL PERMIT FEES (3 YRS)					127,200
TOTAL PWR COSTS					**131,941,455**

(a) GTCC Material: Assumes a low-density, distributed packaging scheme and final disposal as LLW. High-density packaging, ISFSI storage, and geologic repository disposal could reduce disposal costs.

Table B-2. BWR Burial Costs at the Washington Site (2012 dollars)

REFERENCE BWR COMPONENT	VOLUME CHARGE	SHIPMENT CHARGE	CONTAINER CHARGE	CONTAINER DOSE RATE CHARGE	DISPOSAL COST
STEAM SEPARATOR	32,631	192,500	211,680	777,649	1,214,460
FUEL SUPPORT & PIECES	16,315	96,250	105,840	0	218,405
CONTROL RODS/INCORES	48,946	110,000	60,480	0	219,426
CONTROL RODS GUIDES	13,052	82,500	90,720	0	186,272
JET PUMPS	45,683	275,000	302,400	0	623,083
TOP FUEL GUIDES	78,314	990,000	544,320	0	1,612,634
CORE SUPPORT PLATE	35,894	220,000	234,360	0	490,254
CORE SHROUD(a)	153,364	1,925,000	1,058,400	0	3,136,764
REACTOR VESSEL WALL	26,105	275,000	166,320	0	467,425
SAC SHIELD Neutron-Activated Matl	293,676	192,500	105,840	0	592,016
REACT. WATER REC	287,150	68,750	45,360	0	401,260
SAC SHIELD Contaminated Matl	1,011,551	522,500	287,280	0	1,821,331
OTHER PRIMARY CONTAINMENT	11,538,212	2,337,500	7,333,200	0	21,208,912
CONTAINM. ATMOSPHERIC	156,627	13,750	15,120	0	185,497
HIGH PRESSURE CORE SPRAY	55,472	27,500	15,120	0	98,092
LOW PRESSURE CORE SPRAY	32,631	13,750	7,560	0	53,941
REACTOR BLDG CLOSED COOLING	104,418	27,500	45,360	0	177,278
REACTOR CORE ISO COOLING	42,420	13,750	22,680	0	78,850
RESIDUAL HEAT REMOVAL	202,310	68,750	52,920	0	323,980
POOL LINER & RACKS	1,243,229	247,500	279,720	0	1,770,449
CONTAMINATED CONCRETE	1,416,172	385,000	816,480	0	2,617,652
OTHER REACTOR BUILDING	4,630,295	632,500	2,948,400	0	8,211,195
TURBINE	4,587,875	1,127,500	2,101,680	0	7,817,055
NUCLEAR STEAM CONDENSATE	1,184,494	178,750	332,640	0	1,695,884
LOW PRESSURE FEEDWATER HEATERS	2,404,882	577,500	332,640	0	3,315,022
MAIN STEAM	231,678	27,500	22,680	0	281,858
MOISTURE SEPARATOR REHEATERS	2,333,094	357,500	196,560	0	2,887,154
REACTOR FEEDWATER PUMPS	633,035	82,500	151,200	0	866,735
HIGH PRESSURE FEEDWATER HEATERS	394,831	110,000	60,480	0	565,311
OTHER TG BLDG	15,848,726	3,272,500	9,707,040	0	28,828,266
RAD WASTE BLDG	7,847,681	990,000	4,853,520	0	13,691,201
REACTOR BLDG	978,921	522,500	10,795,680	0	12,297,101
TG BLDG	662,403	343,750	7,287,840	0	8,293,993
RAD WASTE & CONTROL	571,037	316,250	6,289,920	0	7,177,207
CONCENTRATOR BOTTOMS	2,088,364	3,093,750	1,701,000	0	6,883,114
OTHER	567,774	838,750	461,160	0	1,867,684
POST-TMI-2 ADDITIONS	117,533	0	0	0	117,533
HEAVY OBJECT SURCHARGE					247,650
SITE AVAILABILITY CHARGES					399,078
SUBTOTAL BWR COSTS	**61,916,797**	**20,556,250**	**59,043,600**	**777,649**	**142,941,025**
TAXES & FEES (% OF CHARGES)					6,146,464
TAXES & FEES ($/UNIT VOL.)					11,559,142
ANNUAL PERMIT FEES (3 YRS)					127,200
TOTAL BWR COSTS					**160,773,831**

(a) GTCC Material: Assumes a low-density, distributed packaging scheme and final disposal as LLW. High-density packaging, ISFSI storage, and geologic repository disposal could reduce disposal costs.

Table B-3. PWR Burial Costs at the Washington Site (2010 dollars)

REFERENCE PWR COMPONENT	VOLUME CHARGE	SHIPMENT CHARGE	CONTAINER CHARGE	CONTAINER DOSE RATE CHARGE	DISPOSAL COST
VESSEL WALL	386,384	508,060	340,480	277,400	1,512,324
VESSEL HEAD & BOTTOM	406,720	534,800	358,400	680	1,300,600
UPPER CORE SUPPORT ASSM	40,672	53,480	35,840	19,400	149,392
UPPER SUPPORT COLUMN	40,672	53,480	35,840	19,400	149,392
UPPER CORE BARREL	20,336	26,740	17,920	14,600	79,596
UPPER CORE GRID PLATE	50,840	66,850	44,800	36,500	198,990
GUIDE TUBES	61,008	80,220	53,760	29,100	224,088
LOWER CORE BARREL[a]	325,376	427,840	286,720	233,600	1,273,536
THERMAL SHIELDS[a]	61,008	80,220	53,760	43,800	238,788
CORE SHROUD[a]	40,672	53,480	35,840	29,200	159,192
LOWER GRID PLATE[a]	50,840	66,850	44,800	36,500	198,990
LOWER SUPPORT COLUMN	10,168	13,370	8,960	7,300	39,798
LOWER CORE FORGING	111,848	147,070	98,560	7,248	364,726
MISC. INTERNALS	81,344	106,960	71,680	0	259,984
BIO SHIELD CONCRETE	2,537,933	655,130	1,747,200	0	4,940,263
REACTOR CAVITY LINER	52,060	13,370	35,840	0	101,270
REACTOR COOLANT PUMPS	427,056	160,440	107,520	0	695,016
PRESSURIZER	366,048	106,960	71,680	0	544,688
R.Hx, EHx, SUMP PUMP, CAVITY PUMP	40,672	13,370	26,880	0	80,922
PRESSURIZER RELIEF TANK	122,016	26,740	17,920	0	166,676
SAFETY INJECTION ACCUM. TANKS	406,720	106,960	71,680	0	585,360
STEAM GENERATORS	2,172,088	427,840	286,720	0	2,886,648
REACTOR COOLANT PIPING	335,544	93,590	62,720	0	491,854
REMAINING CONTAM. MATLS	5,349,181	1,350,370	3,682,560	0	10,382,111
CONTAMINATED MATL OTHER BLDG	48,512,646	10,642,520	33,232,640	0	92,387,806
FILTER CARTRIDGES	32,029	80,220	53,760	0	166,009
SPENT RESINS	203,360	267,400	179,200	0	649,960
COMBUSTIBLE WASTES	1,029,510	802,200	537,600	0	2,369,310
EVAPORATOR BOTTOMS	955,792	1,256,780	842,240	0	3,054,812
POST-TMI-2 ADDITIONS	1,582,446	0	0	0	1,582,446
HEAVY OBJECT SURCHARGE					162,115
SITE AVAILABILITY CHARGES					387,039
SUBTOTAL PWR COSTS	**65,812,990**	**18,223,310**	**42,443,520**	**754,728**	**127,783,702**
TAXES & FEES (% OF CHARGES)					
TAXES & FEES ($/UNIT VOL.)					
ANNUAL PERMIT FEES (3 YRS)					
TOTAL PWR COSTS					**144,536,267**

(a) GTCC Material: Assumes a low-density, distributed packaging scheme and final disposal as LLW. High-density packaging, ISFSI storage, and geologic repository disposal could reduce disposal costs.

Table B-4. BWR Burial Costs at the Washington Site (2010 dollars)

REFERENCE BWR COMPONENT	VOLUME CHARGE	SHIPMENT CHARGE	CONTAINER CHARGE	CONTAINER DOSE RATE CHARGE	DISPOSAL COST
STEAM SEPARATOR	35,908	187,180	250,880	754,728	1,228,696
FUEL SUPPORT & PIECES	17,954	93,590	125,440	0	236,984
CONTROL RODS/INCORES	53,862	106,960	71,680	0	232,502
CONTROL RODS GUIDES	14,363	80,220	107,520	0	202,103
JET PUMPS	50,271	267,400	358,400	0	676,071
TOP FUEL GUIDES	86,179	962,640	645,120	0	1,693,939
CORE SUPPORT PLATE	39,499	213,920	277,760	0	531,179
CORE SHROUD(a)	168,767	1,871,800	1,254,400	0	3,294,967
REACTOR VESSEL WALL	28,726	267,400	197,120	0	493,246
SAC SHIELD NEUTRON ACTIV. MATL	323,171	187,180	125,440	0	635,791
REACTOR WATER REC	315,989	66,850	53,760	0	436,599
SAC SHIELD-CONTAMINATED MATL	1,113,144	508,060	340,480	0	1,961,684
OTHER PRIMARY CONTAINMENT	12,697,028	2,272,900	8,691,200	0	23,661,128
CONTAINMENT. ATMOSPHERIC	172,358	13,370	17,920	0	203,648
HIGH PRESSURE CORE SPRAY	61,043	26,740	17,920	0	105,703
LOW PRESSURE CORE SPRAY	35,908	13,370	8,960	0	58,238
REACTOR BLDG CLOSED COOLING	114,905	26,740	53,760	0	195,405
REACTOR CORE ISO COOLING	46,680	13,370	26,880	0	86,930
RESIDUAL HEAT REMOVAL	222,629	66,850	62,720	0	352,199
POOL LINER & RACKS	1,368,090	240,660	331,520	0	1,940,270
CONTAMINATED CONCRETE	1,558,402	374,360	967,680	0	2,900,442
OTHER REACTOR BUILDING	5,095,329	615,020	3,494,400	0	9,204,749
TURBINE	5,048,649	1,096,340	2,490,880	0	8,635,869
NUCLEAR STEAM CONDENSATE	1,303,456	173,810	394,240	0	1,871,506
LOW PRESSURE FEEDWATER HEATERS	2,646,411	561,540	394,240	0	3,602,191
MAIN STEAM	254,946	26,740	26,880	0	308,566
MOISTURE SEPARATOR REHEATERS	2,567,414	347,620	232,960	0	3,147,994
REACTOR FEEDWATER PUMPS	696,613	80,220	179,200	0	956,033
HIGH PRESSURE FEEDWATER HEATERS	434,485	106,960	71,680	0	613,125
OTHER TG BLDG	17,440,460	3,182,060	11,504,640	0	32,127,160
RAD WASTE BLDG	8,635,846	962,640	5,752,320	0	15,350,806
REACTOR BLDG	1,077,237	508,060	12,794,880	0	14,380,177
TG BLDG	728,930	334,250	8,637,440	0	9,700,620
RAD WASTE & CONTROL	628,388	307,510	7,454,720	0	8,390,618
CONCENTRATOR BOTTOMS	2,298,105	3,008,250	2,016,000	0	7,322,355
OTHER	624,797	815,570	546,560	0	1,986,927
POST-TMI-2 ADDITIONS	129,337	0	0	0	129,337
HEAVY OBJECT SURCHARGE					233,434
SITE AVAILABILITY CHARGES					387,039
SUBTOTAL BWR COSTS	**68,135,281**	**19,988,150**	**69,977,600**	**754,728**	**159,476,232**
TAXES & FEES (% OF CHARGES)					6,857,478
TAXES & FEES ($/UNIT VOL.)					11,559,142
ANNUAL PERMIT FEES (3 YRS)					127,200
TOTAL BWR COSTS					**178,020,053**

(a) GTCC Material: Assumes a low-density, distributed packaging scheme and final disposal as LLW. High-density packaging, ISFSI storage, and geologic repository disposal could reduce disposal costs.

Table B-5. PWR Burial Costs at the Washington Site (2008 dollars)

REFERENCE PWR COMPONENT	VOLUME CHARGE	SHIPMENT CHARGE	CONTAINER CHARGE	CONTAINER DOSE RATE CHARGE	DISPOSAL COST
VESSEL WALL	375,060	560,120	269,040	2,869,000	4,073,220
VESSEL HEAD & BOTTOM	394,800	589,600	283,200	7,080	1,274,680
UPPER CORE SUPPORT ASSM	39,480	58,960	28,320	201,600	328,360
UPPER SUPPORT COLUMN	39,480	58,960	28,320	201,600	328,360
UPPER CORE BARREL	19,740	29,480	14,160	151,000	214,380
UPPER CORE GRID PLATE	49,350	73,700	35,400	377,500	535,950
GUIDE TUBES	59,220	88,440	42,480	302,400	492,540
LOWER CORE BARREL[a]	315,840	471,680	226,560	2,416,000	3,430,080
THERMAL SHIELDS[a]	59,220	88,440	42,480	453,000	643,140
CORE SHROUD[a]	39,480	58,960	28,320	302,000	428,760
LOWER GRID PLATE[a]	49,350	73,700	35,400	377,500	535,950
LOWER SUPPORT COLUMN	9,870	14,740	7,080	75,500	107,190
LOWER CORE FORGING	108,570	162,140	77,880	830,500	1,179,090
MISC. INTERNALS	78,960	117,920	56,640	604,000	857,520
BIO SHIELD CONCRETE	2,463,552	722,260	1,380,600	34,515	4,600,927
REACTOR CAVITY LINER	50,534	14,740	28,320	708	94,302
REACTOR COOLANT PUMPS	414,540	176,880	84,960	2,124	678,504
PRESSURIZER	355,320	117,920	56,640	1,416	531,296
R.Hx, EHx, SUMP PUMP, CAVITY PUMP	39,480	14,740	21,240	531	75,991
PRESSURIZER RELIEF TANK	118,440	29,480	14,160	354	162,434
SAFETY INJECTION ACCUM. TANKS	394,800	117,920	56,640	1,416	570,776
STEAM GENERATORS	2,108,429	471,680	226,560	5,664	2,812,333
REACTOR COOLANT PIPING	325,710	103,180	49,560	1,239	479,689
REMAINING CONTAM. MATLS	5,192,410	1,488,740	2,909,880	72,747	9,663,777
CONTAMINATED MATL OTHER BLDG	47,089,967	11,733,040	26,259,720	656,493	85,739,220
FILTER CARTRIDGES	31,091	88,440	42,480	302,400	464,411
SPENT RESINS	197,400	294,800	141,600	1,510,000	2,143,800
COMBUSTIBLE WASTES	999,338	884,400	424,800	10,620	2,319,158
EVAPORATOR BOTTOMS	927,780	1,385,560	665,520	2,231,879	5,210,739
POST-TMI-2 ADDITIONS	1,536,068	0	0	0	1,536,068
HEAVY OBJECT SURCHARGE					152,809
SITE AVAILABILITY CHARGES (3 YRS)					374,400
SUBTOTAL PWR COSTS	**63,883,279**	**20,090,620**	**33,537,960**	**14,000,786**	**132,039,854**
TAXES & FEES (% OF CHARGES)					5,677,714
TAXES & FEES ($/UNIT VOL.)					11,165,011
ANNUAL PERMIT FEES (3 YRS)					127,200
TOTAL PWR COSTS					**149,009,778**

(a) GTCC Material: Assumes a low-density, distributed packaging scheme and final disposal as LLW. High-density packaging, ISFSI storage, and geologic repository disposal could reduce disposal costs.

Table B-6. BWR Burial Costs at the Washington Site (2008 dollars)

REFERENCE BWR COMPONENT	VOLUME CHARGE	SHIPMENT CHARGE	CONTAINER CHARGE	CONTAINER DOSE RATE CHARGE	DISPOSAL COST
STEAM SEPARATOR	34,841	206,360	198,240	35,504,000	35,943,441
FUEL SUPPORT & PIECES	17,470	103,180	99,120	1,057,000	1,276,770
CONTROL RODS/INCORES	52,311	117,920	56,640	10,144,000	10,370,871
CONTROL RODS GUIDES	13,917	88,440	84,960	906,000	1,093,317
JET PUMPS	48,857	294,800	283,200	50,720,000	51,346,857
TOP FUEL GUIDES	83,698	1,061,280	509,760	91,296,000	92,950,738
CORE SUPPORT PLATE	38,394	235,840	219,480	2,340,500	2,834,214
CORE SHROUD[a]	163,842	2,063,600	991,200	177,520,000	180,738,642
REACTOR VESSEL WALL	27,932	294,800	155,760	1,661,000	2,139,492
SAC SHIELD (NEUTRON ACT. MATL.)	313,669	206,360	99,120	2,478	621,627
REACTOR WATER REC	306,760	73,700	42,480	1,062	424,002
SAC SHIELD (CONTAM. MATL.)	1,080,568	560,120	269,040	6,726	1,916,454
OTHER PRIMARY CONTAINMENT	12,324,669	2,505,800	6,563,160	164,079	21,557,708
CONTAINMENT ATMOSPHERIC	167,297	14,740	14,160	354	196,551
HIGH PRESSURE CORE SPRAY	59,220	29,480	14,160	354	103,214
LOW PRESSURE CORE SPRAY	34,841	14,740	7,080	177	56,838
REACTOR BLDG CLOSED COOLING	111,531	29,480	42,480	1,062	184,553
REACTOR CORE ISO COOLING	45,303	14,740	21,240	531	81,814
RESIDUAL HEAT REMOVAL	216,153	73,700	49,560	1,239	340,652
POOL LINER & RACKS	1,328,009	265,320	261,960	6,549	1,861,838
CONTAMINATED CONCRETE	1,512,775	412,720	764,640	19,116	2,709,251
OTHER REACTOR BUILDING	4,945,857	678,040	2,761,200	69,030	8,454,127
TURBINE	4,900,652	1,208,680	1,968,240	49,206	8,126,778
NUCLEAR STEAM CONDENSATE	1,265,235	191,620	311,520	7,788	1,776,163
LOW PRESSURE FEEDWATER HEATERS	2,568,766	619,080	311,520	7,788	3,507,154
MAIN STEAM	247,540	29,480	21,240	531	298,791
MOISTURE SEPARATOR REHEATERS	2,492,175	383,240	184,080	4,602	3,064,097
REACTOR FEEDWATER PUMPS	676,194	88,440	141,600	3,540	909,774
HIGH PRESSURE FEEDWATER HEATERS	421,745	117,920	56,640	1,416	597,721
OTHER TG BLDG	16,929,024	3,508,120	9,090,720	227,268	29,755,132
RAD WASTE BLDG	8,382,690	1,061,280	4,545,360	113,634	14,102,964
REACTOR BLDG	1,057,077	560,120	10,110,240	252,756	11,980,193
TG BLDG	713,601	368,500	6,825,120	170,628	8,077,849
RAD WASTE & CONTROL	615,888	339,020	5,890,560	147,264	6,992,732
CONCENTRATOR BOTTOMS	2,220,750	3,316,500	1,593,000	5,296,145	12,426,395
OTHER	602,070	899,140	431,880	246,454	2,179,544
POST-TMI-2 ADDITIONS	125,546	0	0	0	125,546
HEAVY OBJECT SURCHARGE					220,034
SITE AVAILABILITY CHARGES (3.5 YRS)					499,200
SUBTOTAL BWR COSTS	**66,146,865**	**22,036,300**	**54,990,360**	**377,950,277**	**521,843,036**
TAXES & FEES (% OF CHARGES)					22,439,251
TAXES & FEES ($/UNIT VOL.)					11,560,622
ANNUAL PERMIT FEES (3.5 YRS)					169,600
TOTAL BWR COSTS					**556,012,508**

(a) GTCC Material: Assumes a low-density, distributed packaging scheme and final disposal as LLW. High-density packaging, ISFSI storage, and geologic repository disposal could reduce disposal costs.

Table B-7. PWR Burial Costs at the Washington Site (2006 dollars)

REFERENCE PWR COMPONENT	VOLUME CHARGE	SHIPMENT CHARGE	CONTAINER CHARGE	CONTAINER DOSE RATE CHARGE	DISPOSAL COST
VESSEL WALL	330,220	460,180	230,280	1,014,600	2,035,280
VESSEL HEAD & BOTTOM	347,600	484,400	242,400	2,520	1,076,920
UPPER CORE SUPPORT ASSM	34,760	48,440	24,240	71,200	178,640
UPPER SUPPORT COLUMN	34,760	48,440	24,240	71,200	178,640
UPPER CORE BARREL	17,380	24,220	12,120	53,400	107,120
UPPER CORE GRID PLATE	43,450	60,550	30,300	133,500	267,800
GUIDE TUBES	52,140	72,660	36,360	106,800	267,960
LOWER CORE BARREL[a]	278,080	387,520	193,920	854,400	1,713,920
THERMAL SHIELDS[a]	52,140	72,660	36,360	160,200	321,360
CORE SHROUD[a]	34,760	48,440	24,240	106,800	214,240
LOWER GRID PLATE[a]	43,450	60,550	30,300	133,500	267,800
LOWER SUPPORT COLUMN	8,690	12,110	6,060	26,700	53,560
LOWER CORE FORGING	95,590	133,210	66,660	293,700	589,160
MISC. INTERNALS	69,520	96,880	48,480	213,600	428,480
BIO SHIELD CONCRETE	2,169,024	593,390	1,181,700	12,285	3,956,399
REACTOR CAVITY LINER	44,493	12,110	24,240	252	81,095
REACTOR COOLANT PUMPS	364,980	145,320	72,720	756	583,776
PRESSURIZER	312,840	96,880	48,480	504	458,704
R.Hx, EHx, SUMP PUMP, CAVITY PUMP	34,760	12,110	18,180	189	65,239
PRESSURIZER RELIEF TANK	104,280	24,220	12,120	126	140,746
SAFETY INJECTION ACCUM. TANKS	347,600	96,880	48,480	504	493,464
STEAM GENERATORS	1,856,358	387,520	193,920	2,016	2,439,814
REACTOR COOLANT PIPING	286,770	84,770	42,420	441	414,401
REMAINING CONTAM. MATLS	4,571,635	1,223,110	2,490,660	25,893	8,311,298
CONTAMIN. MATRL OTHER BLDG	41,460,164	9,639,560	22,476,540	233,667	73,809,931
FILTER CARTRIDGES	27,374	72,660	36,360	106,800	243,194
SPENT RESINS	173,800	242,200	121,200	534,000	1,071,200
COMBUSTIBLE WASTES	879,863	726,600	363,600	3,780	1,973,843
EVAPORATOR BOTTOMS	816,860	1,138,340	569,640	790,701	3,315,541
POST-TMI-2 ADDITIONS	1,352,425	0	0	0	1,352,425
HEAVY OBJECT SURCHARGE					144,483
SITE AVAILABILITY CHARGES (3 YRS)					401,727
SUBTOTAL PWR COSTS	**56,245,764**	**16,505,930**	**28,706,220**	**4,954,034**	**106,958,158**
TAXES & FEES (% OF) CHARGES)					4,599,201
TAXES & FEES ($/UNIT) VOL.)					11,165,011
ANNUAL PERMIT FEES (3 YRS)					127,200
TOTAL PWR COSTS					**122,849,569**

(a) GTCC Material: Assumes a low-density, distributed packaging scheme and final disposal as LLW. High-density packaging, ISFSI storage, and geologic repository disposal could reduce disposal costs.

Table B-8. BWR Burial Costs at the Washington Site (2006 dollars)

REFERENCE BWR COMPONENT	VOLUME CHARGE	SHIPMENT CHARGE	CONTAINER CHARGE	CONTAINER DOSE RATE CHARGE	DISPOSAL COST
STEAM SEPARATOR	30,676	169,540	169,680	12,555,200	12,925,096
FUEL SUPPORT & PIECES	15,381	84,770	84,840	373,800	558,791
CONTROL RODS/INCORES	46,057	96,880	48,480	3,587,200	3,778,617
CONTROL RODS GUIDES	12,253	72,660	72,720	320,400	478,033
JET PUMPS	43,016	242,200	242,400	17,936,000	18,463,616
TOP FUEL GUIDES	73,691	871,920	436,320	32,284,800	33,666,731
CORE SUPPORT PLATE	33,804	193,760	187,860	827,700	1,243,124
CORE SHROUD[a]	144,254	1,695,400	848,400	62,776,000	65,464,054
REACTOR VESSEL WALL	24,593	242,200	133,320	587,400	987,513
SAC SHIELD (NEUTRON ACTIV. MATL)	276,168	169,540	84,840	882	531,430
REACTOR WATER REC	270,085	60,550	36,360	378	367,373
SAC SHIELD (CONTAM. MATL.)	951,381	460,180	230,280	2,394	1,644,235
OTHER PRIMARY CONTAINMENT	10,851,203	2,058,700	5,617,620	58,401	18,585,924
CONTAINMENT ATMOSPHERIC	147,296	12,110	12,120	126	171,652
HIGH PRESSURE CORE SPRAY	52,140	24,220	12,120	126	88,606
LOW PRESSURE CORE SPRAY	30,676	12,110	6,060	63	48,909
REACTOR BLDG CLOSED COOLING	98,197	24,220	36,360	378	159,155
REACTOR CORE ISO COOLING	39,887	12,110	18,180	189	70,366
RESIDUAL HEAT REMOVAL	190,311	60,550	42,420	441	293,722
POOL LINER & RACKS	1,169,240	217,980	224,220	2,331	1,613,771
CONTAMINATED CONCRETE	1,331,916	339,080	654,480	6,804	2,332,280
OTHER REACTOR BUILDING	4,354,559	557,060	2,363,400	24,570	7,299,589
TURBINE	4,314,759	993,020	1,684,680	17,514	7,009,973
NUCLEAR STEAM CONDENSATE	1,113,971	157,430	266,640	2,772	1,540,813
LOW PRESSURE FEEDWATER HEATERS	2,261,659	508,620	266,640	2,772	3,039,691
MAIN STEAM	217,945	24,220	18,180	189	260,534
MOISTURE SEPARATOR REHEATERS	2,194,225	314,860	157,560	1,638	2,668,283
REACTOR FEEDWATER PUMPS	595,352	72,660	121,200	1,260	790,472
HIGH PRESSURE FEEDWATER HEATERS	371,324	96,880	48,480	504	517,188
OTHER TG BLDG	14,905,088	2,882,180	7,781,040	80,892	25,649,200
RAD WASTE BLDG	7,380,504	871,920	3,890,520	40,446	12,183,390
REACTOR BLDG	930,699	460,180	8,653,680	89,964	10,134,523
TG BLDG	628,287	302,750	5,841,840	60,732	6,833,609
RAD WASTE & CONTROL	542,256	278,530	5,041,920	52,416	5,915,122
CONCENTRATOR BOTTOMS	1,955,250	2,724,750	1,363,500	1,876,335	7,919,835
OTHER	530,090	738,710	369,660	87,766	1,726,226
POST-TMI-2 ADDITIONS	110,537	0	0	0	110,537
HEAVY OBJECT SURCHARGE					207,760
SITE AVAILABILITY CHARGES (3.5 YRS)					535,636
SUBTOTAL BWR COSTS	**58,238,729**	**18,104,450**	**47,068,020**	**133,660,783**	**257,815,378**
TAXES & FEES (% OF) CHARGES)					11,086,061
TAXES & FEES ($/UNIT) VOL.)					11,560,622
ANNUAL PERMIT FEES (3.5) YRS)					169,600
TOTAL BWR COSTS					**280,631,661**

(a) GTCC Material: Assumes a low-density, distributed packaging scheme and final disposal as LLW. High-density packaging, ISFSI storage, and geologic repository disposal could reduce disposal costs.

Table B-9. PWR Burial Costs at the Washington Site (2004 dollars)

REFERENCE PWR COMPONENT	VOLUME CHARGE	SHIPMENT CHARGE	CONTAINER CHARGE	LINER DOSE RATE CHARGE	BENTON COUNTY TAX SURCHARGE	DISPOSAL COST
VESSEL WALL	215,080	373,160	187,340	1,520,000	0	2,295,580
VESSEL HEAD & BOTTOM	226,400	392,800	197,200	3,800	0	820,200
UPPER CORE SUPPORT ASSM	22,640	39,280	19,720	107,200	0	188,840
UPPER SUPPORT COLUMN	22,640	39,280	19,720	107,200	0	188,840
UPPER CORE BARREL	11,320	19,640	9,860	80,000	0	120,820
UPPER CORE GRID PLATE	28,300	49,100	24,650	200,000	0	302,050
GUIDE TUBES	33,960	58,920	29,580	160,800	0	283,260
LOWER CORE BARREL[a]	181,120	314,240	157,760	1,280,000	0	1,933,120
THERMAL SHIELDS[a]	33,960	58,920	29,580	240,000	0	362,460
CORE SHROUD[a]	22,640	39,280	19,720	160,000	0	241,640
LOWER GRID PLATE[a]	28,300	49,100	24,650	200,000	0	302,050
LOWER SUPPORT COLUMN	5,660	9,820	4,930	40,000	0	60,410
LOWER CORE FORGING	62,260	108,020	54,230	440,000	0	664,510
MISC. INTERNALS	45,280	78,560	39,440	320,000	0	483,280
BIO SHIELD CONCRETE	1,412,736	481,180	961,350	0	0	2,855,266
REACTOR CAVITY LINER	28,979	9,820	19,720	0	0	58,519
REACTOR COOLANT PUMPS	237,720	117,840	59,160	0	0	414,720
PRESSURIZER	203,760	78,560	39,440	0	0	321,760
R.Hx, EHx, SUMP PUMP, CAVITY PUMP	22,640	9,820	14,790	0	0	47,250
PRESSURIZER RELIEF TANK	67,920	19,640	9,860	0	0	97,420
SAFETY INJECTION ACCUM. TANKS	226,400	78,560	39,440	0	0	344,400
STEAM GENERATORS	1,209,089	314,240	157,760	0	0	1,681,089
REACTOR COOLANT PIPING	186,780	68,740	34,510	0	0	290,030
REMAINING CONTAM. MATLS	2,977,613	991,820	2,026,230	0	0	5,995,663
CONTAM. MATL OTHER BLDG	27,003,973	7,816,720	18,285,370	0	0	53,106,063
FILTER CARTRIDGES	17,829	58,920	29,580	1,125,600	0	1,231,929
SPENT RESINS	113,200	196,400	98,600	800,000	0	1,208,200
COMBUSTIBLE WASTES	573,075	589,200	295,800	0	0	1,458,075
EVAPORATOR BOTTOMS	532,040	923,080	463,420	1,186,315	0	3,104,855
POST-TMI-2 ADDITIONS	880,866	0	0	0	0	880,866
HEAVY OBJECT SURCHARGE						136,313
SITE AVAILABILITY CHARGES (3 YRS)						382,821
SUBTOTAL PWR COSTS	**36,634,180**	**13,384,660**	**23,353,410**	**7,970,915**	**0**	**81,862,299**
TAXES & FEES (% OF CHARGES)						3,520,079
TAXES & FEES ($/UNIT VOL.)						11,165,011
ANNUAL PERMIT FEES (3 YRS)						127,200
TOTAL PWR COSTS						**96,674,588**

(a) GTCC Material: Assumes a low-density, distributed packaging scheme and final disposal as LLW. High-density packaging, ISFSI storage, and geologic repository disposal could reduce disposal costs.

Table B-10. BWR Burial Costs at the Washington Site (2004 dollars)

REFERENCE BWR COMPONENT	VOLUME CHARGE	SHIPMENT CHARGE	CONTAINER CHARGE	LINER DOSE RATE CHARGE	BENTON COUNTY TAX SURCHARGE	DISPOSAL COST
STEAM SEPARATOR	19,980	137,480	138,040	18,816,000	0	19,111,500
FUEL SUPPORT & PIECES	10,018	68,740	69,020	560,000	0	707,778
CONTROL RODS/INCORES	29,998	78,560	39,440	5,376,000	0	5,523,998
CONTROL RODS GUIDES	7,981	58,920	59,160	480,000	0	606,061
JET PUMPS	28,017	196,400	197,200	26,880,000	0	27,301,617
TOP FUEL GUIDES	47,997	707,040	354,960	48,384,000	0	49,493,997
CORE SUPPORT PLATE	22,017	157,120	152,830	1,240,000	0	1,571,967
CORE SHROUD[a]	93,956	1,374,800	690,200	94,080,000	0	96,238,956
REACTOR VESSEL WALL	16,018	196,400	108,460	880,000	0	1,200,878
SAC SHIELD (NEUTRON ACTIV. MATL)	179,875	137,480	69,020	0	0	386,375
REACTOR WATER REC	175,913	49,100	29,580	0	0	254,593
SAC SHIELD (CONTAM. MATL)	619,657	373,160	187,340	0	0	1,180,157
OTHER PRIMARY CONTAINMENT	7,067,642	1,669,400	4,570,110	0	0	13,307,152
CONTAINMENT ATMOSPHERIC	95,937	9,820	9,860	0	0	115,617
HIGH PRESSURE CORE SPRAY	33,960	19,640	9,860	0	0	63,460
LOW PRESSURE CORE SPRAY	19,980	9,820	4,930	0	0	34,730
REACTOR BLDG CLOSED COOLING	63,958	19,640	29,580	0	0	113,178
REACTOR CORE ISO COOLING	25,979	9,820	14,790	0	0	50,589
RESIDUAL HEAT REMOVAL	123,954	49,100	34,510	0	0	207,564
POOL LINER & RACKS	761,553	176,760	182,410	0	0	1,120,723
CONTAMINATED CONCRETE	867,508	274,960	532,440	0	0	1,674,908
OTHER REACTOR BUILDING	2,836,226	451,720	1,922,700	0	0	5,210,646
TURBINE	2,810,303	805,240	1,370,540	0	0	4,986,083
NUCLEAR STEAM CONDENSATE	725,555	127,660	216,920	0	0	1,070,135
LOW PRESSURE FEEDWATER HEATERS	1,473,072	412,440	216,920	0	0	2,102,432
MAIN STEAM	141,953	19,640	14,790	0	0	176,383
MOISTURE SEPARATOR REHEATERS	1,429,150	255,320	128,180	0	0	1,812,650
REACTOR FEEDWATER PUMPS	387,767	58,920	98,600	0	0	545,287
HIGH PRESSURE FEEDWATER HEATERS	241,852	78,560	39,440	0	0	359,852
OTHER TG BLDG	9,708,032	2,337,160	6,330,120	0	0	18,375,312
RAD WASTE BLDG	4,807,095	707,040	3,165,060	0	0	8,679,195
REACTOR BLDG	606,186	373,160	7,040,040	0	0	8,019,386
TG BLDG	409,218	245,500	4,752,520	0	0	5,407,238
RAD WASTE & CONTROL	353,184	225,860	4,101,760	0	0	4,680,804
CONCENTRATOR BOTTOMS	1,273,500	2,209,500	1,109,250	2,815,175	0	7,407,425
OTHER	345,260	599,020	300,730	132,240	0	1,377,250
POST-TMI-2 ADDITIONS	71,995	0	0	0	0	71,995
HEAVY OBJECT SURCHARGE						196,250
SITE AVAILABILITY CHARGES (3.5 YRS)						510,428
SUBTOTAL BWR COSTS	**37,932,245**	**14,680,900**	**38,291,310**	**199,643,415**	**0**	**291,254,548**
TAXES & FEES (% OF) CHARGES)						12,523,946
TAXES & FEES ($/UNIT) VOL.						11,560,622
ANNUAL PERMIT FEES (3.5 YRS)						169,600
TOTAL BWR COSTS						**315,508,715**

(a) GTCC Material: Assumes a low-density, distributed packaging scheme and final disposal as LLW. High-density packaging, ISFSI storage, and geologic repository disposal could reduce disposal costs.

Table B-11. PWR Burial Costs at the Washington Site (2002 dollars)

REFERENCE PWR COMPONENT	VOLUME CHARGE	SHIPMENT CHARGE	CONTAINER CHARGE	LINER DOSE RATE CHARGE	BENTON COUNTY TAX SURCHARGE	DISPOSAL COST
VESSEL WALL	144,020	228,342	78,280	2,101,400	0	2,552,042
VESSEL HEAD & BOTTOM	151,600	240,360	82,400	5,200	0	479,560
UPPER CORE SUPPORT ASSM	15,160	24,036	8,240	147,200	0	194,636
UPPER SUPPORT COLUMN	15,160	24,036	8,240	147,200	0	194,636
UPPER CORE BARREL	7,580	12,018	4,120	110,600	0	134,318
UPPER CORE GRID PLATE	18,950	30,045	10,300	276,500	0	335,795
GUIDE TUBES	22,740	36,054	12,360	220,800	0	291,954
LOWER CORE BARREL[a]	121,280	192,288	65,920	1,769,600	0	2,149,088
THERMAL SHIELDS[a]	22,740	36,054	12,360	331,800	0	402,954
CORE SHROUD[a]	15,160	24,036	8,240	221,200	0	268,636
LOWER GRID PLATE[a]	18,950	30,045	10,300	276,500	0	335,795
LOWER SUPPORT COLUMN	3,790	6,009	2,060	55,300	0	67,159
LOWER CORE FORGING	41,690	66,099	22,660	608,300	0	738,749
MISC. INTERNALS	30,320	48,072	16,480	442,400	0	537,272
BIO SHIELD CONCRETE	945,984	294,441	401,700	0	0	1,642,125
REACTOR CAVITY LINER	19,405	6,009	8,240	0	0	33,654
REACTOR COOLANT PUMPS	159,180	72,108	24,720	0	0	256,008
PRESSURIZER	136,440	48,072	16,480	0	0	200,992
R.Hx, EHx, SUMP PUMP, CAVITY PUMP	15,160	6,009	6,180	0	0	27,349
PRESSURIZER RELIEF TANK	45,480	12,018	4,120	0	0	61,618
SAFETY INJECTION ACCUM. TANKS	151,600	48,072	16,480	0	0	216,152
STEAM GENERATORS	809,620	192,288	65,920	0	0	1,067,828
REACTOR COOLANT PIPING	125,070	42,063	14,420	0	0	181,553
REMAINING CONTAM. MATLS	1,993,843	606,909	846,660	0	0	3,447,412
CONTAM. MATL OTHER BLDG	18,082,166	4,783,164	7,640,540	0	0	30,505,870
FILTER CARTRIDGES	11,939	36,054	12,360	1,545,600	0	1,605,953
SPENT RESINS	75,800	120,180	41,200	1,106,000	0	1,343,180
COMBUSTIBLE WASTES	383,738	360,540	123,600	0	0	867,878
EVAPORATOR BOTTOMS	356,260	564,846	193,640	1,635,910	0	2,750,656
POST-TMI-2 ADDITIONS	589,838	0	0	0	0	589,838
HEAVY OBJECT SURCHARGE						127,975
SITE AVAILABILITY CHARGES (3 YRS)						372,474
SUBTOTAL PWR COSTS	24,530,661	8,190,267	9,758,220	11,001,510	0	53,981,107
TAXES & FEES (% OF CHARGES)						2,051,282
TAXES & FEES ($/UNIT VOL.)						9,223,270
ANNUAL PERMIT FEES (3 YRS)						123,300
TOTAL PWR COSTS						65,378,959

(a) GTCC Material: Assumes a low-density, distributed packaging scheme and final disposal as LLW. High-density packaging, ISFSI storage, and geologic repository disposal could reduce disposal costs.

Table B-12. BWR Burial Costs at the Washington Site (2002 dollars)

REFERENCE BWR COMPONENT	VOLUME CHARGE	SHIPMENT CHARGE	CONTAINER CHARGE	LINER DOSE RATE CHARGE	BENTON COUNTY TAX SURCHARGE	DISPOSAL COST
STEAM SEPARATOR	13,379	84,126	57,680	25,984,000	0	26,139,185
FUEL SUPPORT & PIECES	6,708	42,063	28,840	774,200	0	851,811
CONTROL RODS/INCORES	20,087	48,072	16,480	7,424,000	0	7,508,639
CONTROL RODS GUIDES	5,344	36,054	24,720	663,600	0	729,718
JET PUMPS	18,761	120,180	82,400	37,120,000	0	37,341,341
TOP FUEL GUIDES	32,139	432,648	148,320	66,816,000	0	67,429,107
CORE SUPPORT PLATE	14,743	96,144	63,860	1,714,300	0	1,889,047
CORE SHROUD[(a)]	62,914	841,260	288,400	129,920,000	0	131,112,574
REACTOR VESSEL WALL	10,726	120,180	45,320	1,216,600	0	1,392,826
SAC SHIELD (NEUTRON ACTIV. MATL)	120,446	84,126	28,840	0	0	233,412
REACTOR WATER REC	117,793	30,045	12,360	0	0	160,198
SAC SHIELD (CONTAM. MATL)	414,929	228,342	78,280	0	0	721,551
OTHER PRIMARY CONTAINMENT	4,732,573	1,021,530	1,909,620	0	0	7,663,723
CONTAINMENT ATMOSPHERIC	64,241	6,009	4,120	0	0	74,370
HIGH PRESSURE CORE SPRAY	22,740	12,018	4,120	0	0	38,878
LOW PRESSURE CORE SPRAY	13,379	6,009	2,060	0	0	21,448
REACTOR BLDG CLOSED COOLING	42,827	12,018	12,360	0	0	67,205
REACTOR CORE ISO COOLING	17,396	6,009	6,180	0	0	29,585
RESIDUAL HEAT REMOVAL	83,001	30,045	14,420	0	0	127,466
POOL LINER & RACKS	509,945	108,162	76,220	0	0	694,327
CONTAMINATED CONCRETE	580,893	168,252	222,480	0	0	971,625
OTHER REACTOR BUILDING	1,899,169	276,414	803,400	0	0	2,978,983
TURBINE	1,881,811	492,738	572,680	0	0	2,947,229
NUCLEAR STEAM CONDENSATE	485,840	78,117	90,640	0	0	654,597
LOW PRESSURE FEEDWATER HEATERS	986,385	252,378	90,640	0	0	1,329,403
MAIN STEAM	95,053	12,018	6,180	0	0	113,251
MOISTURE SEPARATOR REHEATERS	956,975	156,234	53,560	0	0	1,166,769
REACTOR FEEDWATER PUMPS	259,653	36,054	41,200	0	0	336,907
HIGH PRESSURE FEEDWATER HEATERS	161,947	48,072	16,480	0	0	226,499
OTHER TG BLDG	6,500,608	1,430,142	2,645,040	0	0	10,575,790
RAD WASTE BLDG	3,218,885	432,648	1,322,520	0	0	4,974,053
REACTOR BLDG	405,909	228,342	2,941,680	0	0	3,575,931
TG BLDG	274,017	150,225	1,985,840	0	0	2,410,082
RAD WASTE & CONTROL	236,496	138,207	1,713,920	0	0	2,088,623
CONCENTRATOR BOTTOMS	852,750	1,352,025	463,500	3,881,970	0	6,550,245
OTHER	231,190	366,549	125,660	181,020	0	904,419
POST-TMI-2 ADDITIONS	48,209	0	0	0	0	48,209
HEAVY OBJECT SURCHARGE						184,275
SITE AVAILABILITY CHARGES (3.5 YRS)						496,632
SUBTOTAL BWR COSTS	**25,399,860**	**8,983,455**	**16,000,020**	**275,695,690**	**0**	**326,759,932**
TAXES & FEES (% OF CHARGES)						12,416,877
TAXES & FEES ($/UNIT VOL.)						9,550,079
ANNUAL PERMIT FEES (3.5 YRS)						164,400
TOTAL BWR COSTS						**348,891,289**

(a) GTCC Material: Assumes a low-density, distributed packaging scheme and final disposal as LLW. High-density packaging, ISFSI storage, and geologic repository disposal could reduce disposal costs.

Table B-13. PWR Burial Costs at the South Carolina Site Atlantic Compact (2012 dollars)

REFERENCE PWR COMPONENT	BASE DISPOSAL CHARGE	CASK HANDLING	CURIE SURCHARGE	LINER DOSE RATE	DOSE RATE SURCHARGE	DISPOSAL COST
VESSEL WALL	4,470,517	3,246,036	8,553,154	0	2,145,848	18,415,555
VESSEL HEAD & BOTTOM	2,848,544	3,416,880	11,260	0	0	6,276,684
UPPER CORE SUPPORT ASSM	268,922	341,688	5,630	0	86,055	702,295
UPPER SUPPORT COLUMN	248,626	341,688	56,300	0	79,560	726,174
UPPER CORE BARREL	118,395	170,844	450,166	0	56,830	796,235
UPPER CORE GRID PLATE	295,988	427,110	1,125,415	0	142,074	1,990,587
GUIDE TUBES	438,062	512,532	56,300	0	118,277	1,125,170
LOWER CORE BARREL[a]	1,894,320	2,733,504	7,202,656	0	909,274	12,739,754
THERMAL SHIELDS[a]	355,185	512,532	1,350,498	0	170,489	2,388,704
CORE SHROUD[a]	275,011	341,688	13,730,063	0	132,005	14,478,767
LOWER GRID PLATE[a]	295,988	427,110	2,250,830	0	142,074	3,116,002
LOWER SUPPORT COLUMN	75,095	85,422	225,083	0	36,046	421,646
LOWER CORE FORGING	815,899	939,642	1,407,500	0	391,632	3,554,673
MISC INTERNALS	661,440	683,376	1,126,000	0	317,491	2,788,307
BIO SHIELD CONCRETE	16,122,600	0	675,600	0	0	16,798,200
REACTOR CAVITY LINER	324,736	0	5,630	0	0	330,366
REACTOR COOLANT PUMPS	5,652,032	0	43,723	0	0	5,695,755
PRESSURIZER	2,565,225	0	2,854	0	0	2,568,079
R.Hx,EHx,SUMP PUMP, CAVITY PUMP	236,790	0	6,638	0	0	243,428
PRESSURIZER RELIEF TANK	710,370	0	2,275	0	0	712,645
SAFETY INJECTION ACCUM TANKS	2,530,008	0	45,851	0	0	2,575,859
STEAM GENERATORS	20,684,032	0	2,477,200	0	0	23,161,232
REACTOR COOLANT PIPING	2,252,856	0	167,774	0	0	2,420,630
REMAINING CONTAM. MATLS	39,825,826	0	125,994	0	0	39,951,820
CONTAMINATED MATRL OTHR BLD	306,381,414	0	103,733	0	0	306,485,147
FILTER CARTRIDGES	405,864	512,532	2,815,000	0	48,704	3,782,100
SPENT RESINS	1,488,240	1,708,440	4,501,660	0	714,355	8,412,695
COMBUSTIBLE WASTES	7,103,700	5,125,320	168,900	0	0	12,397,920
EVAPORATOR BOTTOMS	6,994,728	8,029,668	21,157,802	0	955,450	37,137,648
POST-TMI-2 ADDITIONS	14,036,581	0	0	0	0	14,036,581
SUBTOTAL PWR COSTS	**440,376,992**	**29,556,012**	**69,851,488**	**0**	**6,446,163**	**546,230,654**
ATLANTIC COMPACT COMMISSION ADMINISTRATIVE SURCHARGE						3,883,536
TOTAL PWR COSTS (INSIDE COMPACT)						**550,114,190**

(a) GTCC Material: Assumes a low-density, distributed packaging scheme and final disposal as LLW. High-density packaging, ISFSI storage, and geologic repository disposal could reduce disposal costs.

Table B-14. BWR Burial Costs at the South Carolina Site Atlantic Compact (2012 dollars)

REFERENCE BWR COMPONENT	BASE DISPOSAL CHARGE	CASK HANDLING	CURIE SURCHARGE	LINER DOSE RATE	DOSE RATE SURCHARGE	DISPOSAL COST
STEAM SEPARATOR	275,517	2,391,816	3,151,162	0	132,248	5,950,743
FUEL SUPPORT & PIECES	121,259	1,195,908	394,100	0	58,204	1,769,471
CONTROL RODS/INCORES	361,093	683,376	1,800,664	0	173,325	3,018,458
CONTROL RODS GUIDES	101,506	1,025,064	56,300	0	37,557	1,220,428
JET PUMPS	292,675	3,416,880	4,501,660	0	140,484	8,351,699
TOP FUEL GUIDES	501,729	6,150,384	16,205,976	0	240,830	23,098,918
CORE SUPPORT PLATE	337,215	2,648,082	365,950	0	124,769	3,476,016
CORE SHROUD(a)	982,552	11,959,080	31,511,620	0	471,625	44,924,877
REACTOR VESSEL WALL	213,878	1,879,284	1,216,080	0	79,135	3,388,377
SAC SHIELD Neutron-Activated Matl	4,507,025	0	95,710	0	0	4,602,735
REACT. WATER REC	1,950,435	0	24,745	0	0	1,975,180
SAC SHIELD Contaminated Matl	11,671,868	0	87,169	0	0	11,759,037
OTHER PRIMARY CONTAINMENT	81,872,111	0	994,286	0	0	82,866,397
CONTAINM. ATMOSPHERIC	1,003,457	0	13,497	0	0	1,016,954
HIGH PRESSURE CORE SPRAY	507,463	0	4,780	0	0	512,243
LOW PRESSURE CORE SPRAY	225,067	0	2,812	0	0	227,879
REACTOR BLDG CLOSED COOLING	788,849	0	8,998	0	0	797,847
REACTOR CORE ISO COOLING	271,770	0	3,655	0	0	275,425
RESIDUAL HEAT REMOVAL	1,499,238	0	17,434	0	0	1,516,672
POOL LINER & RACKS	10,054,928	0	107,133	0	0	10,162,061
CONTAMINATED CONCRETE	11,090,053	0	122,036	0	0	11,212,090
OTHER REACTOR BUILDING	29,664,706	0	399,008	0	0	30,063,714
TURBINE	37,200,563	0	395,352	0	0	37,595,915
NUCLEAR STEAM CONDENSATE	7,588,646	0	102,072	0	0	7,690,718
LOW PRESSURE FEEDWATER HEATERS	16,510,913	0	207,237	0	0	16,718,149
MAIN STEAM	1,484,281	0	19,964	0	0	1,504,245
MOISTURE SEPARATOR REHEATERS	14,947,333	0	201,050	0	0	15,148,383
REACTOR FEEDWATER PUMPS	4,055,640	0	54,551	0	0	4,110,191
HIGH PRESSURE FEEDWATER HEATERS	2,678,760	0	34,024	0	0	2,712,784
OTHER TG BLDG	101,537,335	0	1,365,737	0	0	102,903,072
RAD WASTE BLDG	50,277,392	0	676,260	0	0	50,953,652
REACTOR BLDG	12,851,317	5,467,008	106,970	0	0	18,425,295
TG BLDG	8,454,814	3,587,724	70,375	0	0	12,112,913
RAD WASTE & CONTROL	7,778,429	3,246,036	64,745	0	0	11,089,210
CONCENTRATOR BOTTOMS	28,744,710	19,219,950	50,643,675	0	3,893,950	102,502,285
OTHER	7,793,010	5,210,742	539,917	0	194,186	13,737,856
POST-TMI-2 ADDITIONS	1,147,242	0	0	0	0	1,147,242
SUBTOTAL BWR COSTS	**461,344,779**	**68,081,334**	**115,566,704**	**0**	**5,546,314**	**650,539,131**
ATLANTIC COMPACT COMMISSION ADMINISTRATIVE SURCHARGE						4,020,571
TOTAL BWR COSTS (INSIDE COMPACT)						**654,559,702**

(a) GTCC Material: Assumes a low-density, distributed packaging scheme and final disposal as LLW. High-density packaging, ISFSI storage, and geologic repository disposal could reduce disposal costs.

Table B-15. PWR Burial Costs at the South Carolina Site Atlantic Compact (2010 dollars)

REFERENCE PWR COMPONENT	BASE DISPOSAL CHARGE	CASK HANDLING	CURIE SURCHARGE	LINER DOSE RATE	DOSE RATE SURCHARGE	DISPOSAL COST
VESSEL WALL	3,986,350	2,894,422	7,626,676	0	1,913,448	16,420,895
VESSEL HEAD & BOTTOM	2,540,054	3,046,760	10,040	0	0	5,596,854
UPPER CORE SUPPORT ASSM	239,799	304,676	5,020	0	76,736	626,230
UPPER SUPPORT COLUMN	221,701	304,676	50,200	0	70,944	647,521
UPPER CORE BARREL	105,570	152,338	401,404	0	50,674	709,986
UPPER CORE GRID PLATE	263,925	380,845	1,003,510	0	126,684	1,774,964
GUIDE TUBES	390,609	457,014	50,200	0	105,464	1,003,287
LOWER CORE BARREL[a]	1,689,120	2,437,408	6,422,464	0	810,778	11,359,770
THERMAL SHIELDS[a]	316,710	457,014	1,204,212	0	152,021	2,129,957
CORE SHROUD[a]	245,228	304,676	12,242,822	0	117,709	12,910,435
LOWER GRID PLATE[a]	263,925	380,845	2,007,020	0	126,684	2,778,474
LOWER SUPPORT COLUMN	66,963	76,169	200,702	0	32,142	375,976
LOWER CORE FORGING	727,540	837,859	1,255,000	0	349,219	3,169,618
MISC. INTERNALS	589,840	609,352	1,004,000	0	283,123	2,486,315
BIO SHIELD CONCRETE	14,377,350	0	602,400	0	0	14,979,750
REACTOR CAVITY LINER	289,568	0	5,020	0	0	294,588
REACTOR COOLANT PUMPS	5,039,904	0	38,985	0	0	5,078,889
PRESSURIZER	2,287,350	0	2,545	0	0	2,289,895
R.Hx, EHx, SUMP PUMP, CAVITY PUMP	211,140	0	5,919	0	0	217,059
PRESSURIZER RELIEF TANK	633,420	0	2,028	0	0	635,448
SAFETY INJECTION ACCUM. TANKS	2,256,138	0	40,883	0	0	2,297,021
STEAM GENERATORS	18,443,904	0	2,208,800	0	0	20,652,704
REACTOR COOLANT PIPING	2,008,878	0	149,596	0	0	2,158,474
REMAINING CONTAM. MATLS	35,512,801	0	112,343	0	0	35,625,143
CONTAM. MATL OTHER BLDG	273,201,164	0	92,494	0	0	273,293,658
FILTER CARTRIDGES	361,908	457,014	2,510,000	0	43,429	3,372,351
SPENT RESINS	1,327,140	1,523,380	4,014,040	0	637,027	7,501,587
COMBUSTIBLE WASTES	6,334,200	4,570,140	150,600	0	0	11,054,940
EVAPORATOR BOTTOMS	6,237,558	7,159,886	18,865,988	0	852,024	33,115,456
POST-TMI-2 ADDITIONS	12,516,387	0	0	0	0	12,516,387
SUBTOTAL PWR COSTS	**392,686,142**	**26,354,474**	**62,284,910**	**0**	**5,748,106**	**487,073,631**
ATLANTIC COMPACT COMMISSION ADMINISTRATIVE SURCHARGE						3,883,536
TOTAL PWR COSTS (INSIDE COMPACT)						**490,957,167**

(a) GTCC Material: Assumes a low-density, distributed packaging scheme and final disposal as LLW. High-density packaging, ISFSI storage, and geologic repository disposal could reduce disposal costs.

Table B-16. BWR Burial Costs at the South Carolina Site Atlantic Compact (2010 dollars)

REFERENCE BWR COMPONENT	BASE DISPOSAL CHARGE	CASK HANDLING	CURIE SURCHARGE	LINER DOSE RATE	DOSE RATE SURCHARGE	DISPOSAL COST
STEAM SEPARATOR	245,672	2,132,732	2,809,828	0	117,923	5,306,155
FUEL SUPPORT & PIECES	108,127	1,066,366	351,400	0	51,901	1,577,794
CONTROL RODS/INCORES	322,005	609,352	1,605,616	0	154,563	2,691,536
CONTROL RODS GUIDES	90,511	914,028	50,200	0	33,489	1,088,228
JET PUMPS	260,971	3,046,760	4,014,040	0	125,266	7,447,038
TOP FUEL GUIDES	447,379	5,484,168	14,450,544	0	214,742	20,596,834
CORE SUPPORT PLATE	300,712	2,361,239	326,300	0	111,263	3,099,514
CORE SHROUD(a)	876,118	10,663,660	28,098,280	0	420,537	40,058,595
REACTOR VESSEL WALL	190,716	1,675,718	1,084,320	0	70,565	3,021,319
SAC SHIELD NEUTRON ACTIV. MATL	4,018,904	0	85,340	0	0	4,104,244
REACTOR WATER REC	1,739,209	0	22,064	0	0	1,761,273
SAC SHIELD-CONTAMINATED MATL	10,407,780	0	77,724	0	0	10,485,504
OTHER PRIMARY CONTAINMENT	73,003,420	0	886,557	0	0	73,889,977
CONTAINMENT ATMOSPHERIC	894,759	0	12,035	0	0	906,794
HIGH PRESSURE CORE SPRAY	452,530	0	4,262	0	0	456,793
LOW PRESSURE CORE SPRAY	200,693	0	2,507	0	0	203,201
REACTOR BLDG CLOSED COOLING	703,398	0	8,023	0	0	711,421
REACTOR CORE ISO COOLING	242,331	0	3,259	0	0	245,590
RESIDUAL HEAT REMOVAL	1,336,948	0	15,545	0	0	1,352,493
POOL LINER & RACKS	8,965,741	0	95,525	0	0	9,061,266
CONTAMINATED CONCRETE	9,889,033	0	108,814	0	0	9,997,847
OTHER REACTOR BUILDING	26,451,312	0	355,776	0	0	26,807,088
TURBINE	33,171,850	0	352,517	0	0	33,524,367
NUCLEAR STEAM CONDENSATE	6,766,615	0	91,012	0	0	6,857,627
LOW PRESSURE FEEDWATER HEATERS	14,722,829	0	184,783	0	0	14,907,612
MAIN STEAM	1,323,498	0	17,801	0	0	1,341,299
MOISTURE SEPARATOR REHEATERS	13,328,180	0	179,267	0	0	13,507,447
REACTOR FEEDWATER PUMPS	3,616,317	0	48,640	0	0	3,664,958
HIGH PRESSURE FEEDWATER HEATERS	2,388,788	0	30,337	0	0	2,419,126
OTHER TG BLDG	90,538,422	0	1,217,762	0	0	91,756,184
RAD WASTE BLDG	44,831,152	0	602,989	0	0	45,434,141
REACTOR BLDG	11,459,491	4,874,816	95,380	0	0	16,429,687
TG BLDG	7,539,139	3,199,098	62,750	0	0	10,800,987
RAD WASTE & CONTROL	6,936,008	2,894,422	57,730	0	0	9,888,160
CONCENTRATOR BOTTOMS	25,631,595	17,138,025	45,157,950	0	3,472,227	91,399,797
OTHER	6,949,010	4,646,309	481,418	0	173,156	12,249,893
POST-TMI-2 ADDITIONS	1,022,993	0	0	0	0	1,022,993
SUBTOTAL BWR COSTS	**411,374,158**	**60,706,693**	**103,048,296**	**0**	**4,945,631**	**580,074,778**
ATLANTIC COMPACT COMMISSION ADMINISTRATIVE SURCHARGE						4,020,571
TOTAL BWR COSTS (INSIDE COMPACT)						**584,095,349**

(a) GTCC Material: Assumes a low-density, distributed packaging scheme and final disposal as LLW. High-density packaging, ISFSI storage, and geologic repository disposal could reduce disposal costs.

NUREG-1307

Table B-17. PWR Burial Costs at the South Carolina Site Atlantic Compact (2008 dollars)

REFERENCE PWR COMPONENT	BASE DISPOSAL CHARGE	CASK HANDLING	CURIE SURCHARGE	LINER DOSE RATE	DOSE RATE SURCHARGE	DISPOSAL COST
VESSEL WALL	3,682,407	2,673,832	7,052,800	0	1,767,555	15,176,594
VESSEL HEAD & BOTTOM	2,346,371	2,814,560	9,280	0	0	5,170,211
UPPER CORE SUPPORT ASSM	221,514	281,456	4,640	0	70,884	578,494
UPPER SUPPORT COLUMN	204,796	281,456	46,400	0	65,535	598,186
UPPER CORE BARREL	97,524	140,728	371,200	0	46,812	656,264
UPPER CORE GRID PLATE	243,810	351,820	928,000	0	117,029	1,640,659
GUIDE TUBES	360,839	422,184	46,400	0	97,426	926,849
LOWER CORE BARREL[a]	1,560,384	2,251,648	5,939,200	0	748,984	10,500,216
THERMAL SHIELDS[a]	292,572	422,184	1,113,600	0	140,435	1,968,791
CORE SHROUD[a]	226,529	281,456	11,321,600	0	108,734	11,938,319
LOWER GRID PLATE[a]	243,810	351,820	1,856,000	0	117,029	2,568,659
LOWER SUPPORT COLUMN	61,857	70,364	185,600	0	29,691	347,512
LOWER CORE FORGING	672,064	774,004	1,160,000	0	322,591	2,928,658
MISC. INTERNALS	544,880	562,912	928,000	0	261,542	2,297,334
BIO SHIELD CONCRETE	13,281,450	0	556,800	0	0	13,838,250
REACTOR CAVITY LINER	267,488	0	4,640	0	0	272,128
REACTOR COOLANT PUMPS	4,655,632	0	36,034	0	0	4,691,666
PRESSURIZER	2,113,020	0	2,352	0	0	2,115,372
R.Hx, EHx, SUMP PUMP, CAVITY PUMP	195,048	0	5,471	0	0	200,519
PRESSURIZER RELIEF TANK	585,144	0	1,875	0	0	587,019
SAFETY INJECTION ACCUM. TANKS	2,084,166	0	37,788	0	0	2,121,954
STEAM GENERATORS	17,037,632	0	2,041,600	0	0	19,079,232
REACTOR COOLANT PIPING	1,855,681	0	138,272	0	0	1,993,953
REMAINING CONTAM. MATLS	32,804,896	0	103,839	0	0	32,908,734
CONTAM. MATL OTHER BLDG	252,369,160	0	85,492	0	0	252,454,652
FILTER CARTRIDGES	334,314	422,184	2,320,000	0	40,118	3,116,616
SPENT RESINS	1,225,980	1,407,280	3,712,000	0	588,470	6,933,730
COMBUSTIBLE WASTES	5,851,440	4,221,840	139,200	0	0	10,212,480
EVAPORATOR BOTTOMS	5,762,106	6,614,216	17,446,400	0	787,079	30,609,801
POST-TMI-2 ADDITIONS	11,562,064	0	0	0	0	11,562,064
ATLANTIC COMPACT COMMISSION ADMINISTRATIVE SURCHARGE						3,883,482
TOTAL PWR COSTS (INSIDE COMPACT)						**453,878,398**

(a) GTCC Material: Assumes a low-density, distributed packaging scheme and final disposal as LLW. High-density packaging, ISFSI storage, and geologic repository disposal could reduce disposal costs.

Table B-18. BWR Burial Costs at the South Carolina Site Atlantic Compact (2008 dollars)

REFERENCE BWR COMPONENT	BASE DISPOSAL CHARGE	CASK HANDLING	CURIE SURCHARGE	LINER DOSE RATE	DOSE RATE SURCHARGE	DISPOSAL COST
STEAM SEPARATOR	226,342	1,970,192	2,598,400	0	108,644	4,903,579
FUEL SUPPORT & PIECES	99,706	985,096	324,800	0	47,859	1,457,461
CONTROL RODS/INCORES	296,851	562,912	1,484,800	0	142,488	2,487,051
CONTROL RODS GUIDES	83,437	844,368	46,400	0	30,872	1,005,077
JET PUMPS	241,372	2,814,560	3,712,000	0	115,859	6,883,790
TOP FUEL GUIDES	413,502	5,066,208	13,363,200	0	198,481	19,041,391
CORE SUPPORT PLATE	277,208	2,181,284	301,600	0	102,567	2,862,659
CORE SHROUD [a]	809,449	9,850,960	25,984,000	0	388,536	37,032,945
REACTOR VESSEL WALL	176,108	1,548,008	1,002,240	0	65,160	2,791,516
SAC SHIELD	3,704,707	0	77,952	0	0	3,782,659
REACTOR WATER REC	1,603,223	0	20,391	0	0	1,623,614
SAC SHIELD	9,594,069	0	71,829	0	0	9,665,898
OTHER PRIMARY CONTAINMENT	67,298,062	0	819,267	0	0	68,117,329
CONTAINMENT ATMOSPHERIC	826,516	0	11,121	0	0	837,637
HIGH PRESSURE CORE SPRAY	415,471	0	3,937	0	0	419,408
LOW PRESSURE CORE SPRAY	183,898	0	2,316	0	0	186,214
REACTOR BLDG CLOSED COOLING	648,426	0	7,414	0	0	655,840
REACTOR CORE ISO COOLING	223,818	0	3,011	0	0	226,829
RESIDUAL HEAT REMOVAL	1,253,238	0	14,369	0	0	1,267,606
POOL LINER & RACKS	8,265,051	0	88,278	0	0	8,353,328
CONTAMINATED CONCRETE	9,115,824	0	100,560	0	0	9,216,384
OTHER REACTOR BUILDING	24,434,638	0	328,770	0	0	24,763,408
TURBINE	30,578,192	0	325,765	0	0	30,903,956
NUCLEAR STEAM CONDENSATE	6,250,801	0	84,105	0	0	6,334,906
LOW PRESSURE FEEDWATER HEATERS	13,571,672	0	170,756	0	0	13,742,428
MAIN STEAM	1,222,951	0	16,455	0	0	1,239,406
MOISTURE SEPARATOR REHEATERS	12,312,405	0	165,664	0	0	12,478,069
REACTOR FEEDWATER PUMPS	3,340,685	0	44,949	0	0	3,385,634
HIGH PRESSURE FEEDWATER HEATERS	2,202,078	0	28,035	0	0	2,230,113
OTHER TG BLDG	83,636,582	0	1,125,336	0	0	84,761,918
RAD WASTE BLDG	41,414,054	0	557,229	0	0	41,971,283
REACTOR BLDG	10,585,738	4,503,296	88,160	0	0	15,177,194
TG BLDG	6,964,301	2,955,288	58,000	0	0	9,977,589
RAD WASTE & CONTROL	6,407,157	2,673,832	53,360	0	0	9,134,349
CONCENTRATOR BOTTOMS	23,677,260	15,831,900	41,760,000	0	3,207,479	84,476,639
OTHER	6,419,168	4,292,204	444,976	0	159,953	11,316,301
POST-TMI-2 ADDITIONS	944,994	0	0	0	0	944,994
SUBTOTAL BWR COSTS	379,718,953	56,080,108	95,289,444	0	4,567,898	535,656,402
ATLANTIC COMPACT COMMISSION ADMINISTRATIVE SURCHARGE						4,021,086
TOTAL BWR COSTS (INSIDE COMPACT)						539,677,488

(a) GTCC Material: Assumes a low-density, distr buted packaging scheme and final disposal as LLW. High-density packaging, ISFSI storage, and geologic repository disposal could reduce disposal costs.

Table B-19. PWR Burial Costs at the South Carolina Site Atlantic Compact (2006 dollars)

REFERENCE PWR COMPONENT	BASE DISPOSAL CHARGE	CASK HANDLING	CURIE SURCHARGE	LINER DOSE RATE	DOSE RATE SURCHARGE	DISPOSAL COST
VESSEL WALL	3,344,560	2,428,580	6,399,200	0	1,605,389	13,777,729
VESSEL HEAD & BOTTOM	2,131,074	2,556,400	8,420	0	0	4,695,894
UPPER CORE SUPPORT ASSM	201,188	255,640	4,210	0	64,380	525,418
UPPER SUPPORT COLUMN	186,004	255,640	42,100	0	59,521	543,265
UPPER CORE BARREL	88,578	127,820	336,800	0	42,517	595,715
UPPER CORE GRID PLATE	221,445	319,550	842,000	0	106,294	1,489,289
GUIDE TUBES	327,739	383,460	42,100	0	88,489	841,788
LOWER CORE BARREL[a]	1,417,248	2,045,120	5,388,800	0	680,279	9,531,447
THERMAL SHIELDS[a]	265,734	383,460	1,010,400	0	127,552	1,787,146
CORE SHROUD[a]	205,743	255,640	10,272,400	0	98,757	10,832,540
LOWER GRID PLATE[a]	221,445	319,550	1,684,000	0	106,294	2,331,289
LOWER SUPPORT COLUMN	56,181	63,910	168,400	0	26,967	315,458
LOWER CORE FORGING	610,397	703,010	1,052,500	0	292,990	2,658,897
MISC. INTERNALS	494,880	511,280	842,000	0	237,542	2,085,702
BIO SHIELD CONCRETE	12,062,700	0	505,200	0	0	12,567,900
REACTOR CAVITY LINER	242,944	0	4,210	0	0	247,154
REACTOR COOLANT PUMPS	4,228,496	0	32,695	0	0	4,261,191
PRESSURIZER	1,919,190	0	2,134	0	0	1,921,324
R.Hx, EHx, SUMP PUMP, CAVITY PUMP	177,156	0	4,964	0	0	182,120
PRESSURIZER RELIEF TANK	531,468	0	1,701	0	0	533,169
SAFETY INJECTION ACCUM. TANKS	1,892,916	0	34,286	0	0	1,927,202
STEAM GENERATORS	15,474,496	0	1,852,400	0	0	17,326,896
REACTOR COOLANT PIPING	1,685,409	0	125,458	0	0	1,810,867
REMAINING CONTAM. MATLS	29,794,804	0	94,216	0	0	29,889,020
CONTAM. MATL OTHER BLDG	229,212,426	0	77,569	0	0	229,289,995
FILTER CARTRIDGES	303,642	383,460	2,105,000	0	36,437	2,828,539
SPENT RESINS	1,113,480	1,278,200	3,368,000	0	534,470	6,294,150
COMBUSTIBLE WASTES	5,314,680	3,834,600	126,300	0	0	9,275,580
EVAPORATOR BOTTOMS	5,233,356	6,007,540	15,829,600	0	714,854	27,785,350
POST-TMI-2 ADDITIONS	10,501,290	0	0	0	0	10,501,290
SUBTOTAL PWR COSTS	**329,460,668**	**22,112,860**	**52,257,063**	**0**	**4,822,734**	**408,653,325**
ATLANTIC COMPACT COMMISSION ADMINISTRATIVE SURCHARGE						3,883,482
TOTAL PWR COSTS (INSIDE COMPACT)						**412,536,807**

(a) GTCC Material: Assumes a low-density, distributed packaging scheme and final disposal as LLW. High-density packaging, ISFSI storage, and geologic repository disposal could reduce disposal costs.

Table B-20. BWR Burial Costs at the South Carolina Site Atlantic Compact (2006 dollars)

REFERENCE BWR COMPONENT	BASE DISPOSAL CHARGE	CASK HANDLING	CURIE SURCHARGE	LINER DOSE RATE	DOSE RATE SURCHARGE	DISPOSAL COST
STEAM SEPARATOR	205,580	1,789,480	2,357,600	0	98,678	4,451,338
FUEL SUPPORT & PIECES	90,557	894,740	294,700	0	43,468	1,323,465
CONTROL RODS/INCORES	269,611	511,280	1,347,200	0	129,413	2,257,504
CONTROL RODS GUIDES	75,783	766,920	42,100	0	28,040	912,843
JET PUMPS	219,231	2,556,400	3,368,000	0	105,231	6,248,861
TOP FUEL GUIDES	375,571	4,601,520	12,124,800	0	180,274	17,282,165
CORE SUPPORT PLATE	251,770	1,981,210	273,650	0	93,155	2,599,785
CORE SHROUD[a]	735,197	8,947,400	23,576,000	0	352,895	33,611,492
REACTOR VESSEL WALL	159,949	1,406,020	909,360	0	59,181	2,534,509
SAC SHIELD	3,364,814	0	70,728	0	0	3,435,542
REACTOR WATER REC	1,456,115	0	18,502	0	0	1,474,617
SAC SHIELD	8,713,851	0	65,173	0	0	8,779,023
OTHER PRIMARY CONTAINMENT	61,124,725	0	743,344	0	0	61,868,069
CONTAINMENT ATMOSPHERIC	750,699	0	10,090	0	0	760,789
HIGH PRESSURE CORE SPRAY	377,346	0	3,572	0	0	380,918
LOW PRESSURE CORE SPRAY	167,024	0	2,101	0	0	169,125
REACTOR BLDG CLOSED COOLING	588,945	0	6,727	0	0	595,672
REACTOR CORE ISO COOLING	203,287	0	2,732	0	0	206,019
RESIDUAL HEAT REMOVAL	1,138,236	0	13,037	0	0	1,151,273
POOL LINER & RACKS	7,506,887	0	80,097	0	0	7,586,984
CONTAMINATED CONCRETE	8,279,380	0	91,241	0	0	8,370,620
OTHER REACTOR BUILDING	22,193,218	0	298,302	0	0	22,491,520
TURBINE	27,772,417	0	295,575	0	0	28,067,992
NUCLEAR STEAM CONDENSATE	5,677,407	0	76,311	0	0	5,753,718
LOW PRESSURE FEEDWATER HEATERS	12,326,371	0	154,931	0	0	12,481,302
MAIN STEAM	1,110,768	0	14,930	0	0	1,125,698
MOISTURE SEPARATOR REHEATERS	11,182,973	0	150,312	0	0	11,333,284
REACTOR FEEDWATER PUMPS	3,034,239	0	40,784	0	0	3,075,023
HIGH PRESSURE FEEDWATER HEATERS	2,000,008	0	25,437	0	0	2,025,445
OTHER TG BLDG	75,964,493	0	1,021,048	0	0	76,985,541
RAD WASTE BLDG	37,615,091	0	505,589	0	0	38,120,680
REACTOR BLDG	9,614,538	4,090,240	79,990	0	0	13,784,768
TG BLDG	6,325,354	2,684,220	52,625	0	0	9,062,199
RAD WASTE & CONTROL	5,819,326	2,428,580	48,415	0	0	8,296,321
CONCENTRATOR BOTTOMS	21,504,964	14,379,750	37,890,000	0	2,913,206	76,687,920
OTHER	5,830,235	3,898,510	403,739	0	145,278	10,277,762
POST-TMI-2 ADDITIONS	858,295	0	0	0	0	858,295
SUBTOTAL BWR COSTS	**344,884,253**	**50,936,270**	**86,458,741**	**0**	**4,148,818**	**486,428,082**
ATLANTIC COMPACT COMMISSION ADMINISTRATIVE SURCHARGE						4,021,086
TOTAL BWR COSTS (INSIDE COMPACT)						**490,449,168**

(a) GTCC Material: Assumes a low-density, distr buted packaging scheme and final disposal as LLW. High-density packaging, ISFSI storage, and geologic repository disposal could reduce disposal costs.

Table B-21. PWR Burial Costs at the South Carolina Site Non-Atlantic Compact (2006 dollars)

REFERENCE PWR COMPONENT	BASE DISPOSAL CHARGE	CASK HANDLING	CURIE SURCHARGE	LINER DOSE RATE	DOSE RATE SURCHARGE	DISPOSAL COST
VESSEL WALL	3,207,162	2,686,790	6,874,200	0	1,539,438	14,307,589
VESSEL HEAD & BOTTOM	1,929,813	2,828,200	9,000	0	0	4,767,013
UPPER CORE SUPPORT ASSM	191,569	282,820	4,500	0	61,302	540,190
UPPER SUPPORT COLUMN	188,724	282,820	45,000	0	60,392	576,935
UPPER CORE BARREL	71,112	141,410	362,000	0	34,134	608,656
UPPER CORE GRID PLATE	163,200	353,525	943,380	0	78,336	1,538,441
GUIDE TUBES	288,112	424,230	45,000	0	77,790	835,132
LOWER CORE BARREL[a]	1,336,868	2,262,560	6,932,032	0	641,697	11,173,157
THERMAL SHIELDS[a]	258,980	424,230	1,340,000	0	124,310	2,147,520
CORE SHROUD[a]	195,906	282,820	11,381,968	0	94,035	11,954,729
LOWER GRID PLATE[a]	185,597	353,525	2,293,800	0	89,087	2,922,009
LOWER SUPPORT COLUMN	50,875	70,705	200,000	0	24,420	346,000
LOWER CORE FORGING	552,750	777,755	1,125,000	0	265,320	2,720,825
MISC. INTERNALS	455,120	565,640	900,000	0	218,458	2,139,218
BIO SHIELD CONCRETE	12,017,850	0	540,000	0	0	12,557,850
REACTOR CAVITY LINER	246,496	0	4,500	0	0	250,996
REACTOR COOLANT PUMPS	4,054,784	0	34,947	0	0	4,089,731
PRESSURIZER	1,848,990	0	2,282	0	0	1,851,272
R.Hx, EHx, SUMP PUMP, CAVITY PUMP	167,820	0	5,306	0	0	173,125
PRESSURIZER RELIEF TANK	435,200	0	1,818	0	0	437,018
SAFETY INJECTION ACCUM. TANKS	1,885,878	0	36,648	0	0	1,922,526
STEAM GENERATORS	15,169,024	0	1,980,000	0	0	17,149,024
REACTOR COOLANT PIPING	1,604,824	0	134,100	0	0	1,738,924
REMAINING CONTAM. MATLS	26,980,938	0	100,706	0	0	27,081,643
CONTAM. MATL OTHER BLDG	232,563,661	0	0	0	0	232,563,661
FILTER CARTRIDGES	291,168	424,230	2,260,000	0	23,293	2,998,691
SPENT RESINS	1,055,880	1,414,100	3,684,000	0	506,822	6,660,802
COMBUSTIBLE WASTES	5,120,280	4,242,300	135,000	0	0	9,497,580
EVAPORATOR BOTTOMS	4,962,636	6,646,270	17,000,528	0	487,817	29,097,251
POST-TMI-2 ADDITIONS	5,754,886	0	0	0	0	5,754,886
SUBTOTAL PWR COSTS	**323,236,099**	**24,463,930**	**58,375,714**	**0**	**4,326,649**	**410,402,392**
ATLANTIC COMPACT COMMISSION ADMINISTRATIVE SURCHARGE						3,883,482
TOTAL PWR COSTS (OUTSIDE COMPACT)						**414,285,874**

(a) GTCC Material: Assumes a low-density, distributed packaging scheme and final disposal as LLW. High-density packaging, ISFSI storage, and geologic repository disposal could reduce disposal costs.

Table B-22. BWR Burial Costs at the South Carolina Site Non-Atlantic Compact (2006 dollars)

REFERENCE BWR COMPONENT	BASE DISPOSAL CHARGE	CASK HANDLING	CURIE SURCHARGE	LINER DOSE RATE	DOSE RATE SURCHARGE	DISPOSAL COST
STEAM SEPARATOR	180,723	1,979,740	2,539,208	0	86,747	4,786,418
FUEL SUPPORT & PIECES	86,228	989,870	315,000	0	41,389	1,432,487
CONTROL RODS/INCORES	263,465	565,640	1,818,000	0	126,463	2,773,569
CONTROL RODS GUIDES	73,011	848,460	45,000	0	27,014	993,486
JET PUMPS	187,730	2,828,200	3,640,000	0	90,111	6,746,041
TOP FUEL GUIDES	331,666	5,090,760	13,020,192	0	159,200	18,601,818
CORE SUPPORT PLATE	231,542	2,191,855	292,500	0	85,671	2,801,568
CORE SHROUD[a]	751,170	9,898,700	37,800,000	0	360,562	48,810,432
REACTOR VESSEL WALL	144,843	1,555,510	972,000	0	53,592	2,725,945
SAC SHIELD	3,226,584	0	75,600	0	0	3,302,184
REACT. WATER REC	1,477,405	0	19,776	0	0	1,497,181
SAC SHIELD	8,355,875	0	69,662	0	0	8,425,537
OTHER PRIMARY CONTAINMENT	62,565,584	0	794,548	0	0	63,360,132
CONTAINM. ATMOSPHERIC	651,850	0	10,785	0	0	662,635
HIGH PRESSURE CORE SPRAY	347,029	0	3,818	0	0	350,847
LOW PRESSURE CORE SPRAY	169,466	0	2,246	0	0	171,712
REACTOR BLDG CLOSED COOLING	567,403	0	7,190	0	0	574,593
REACTOR CORE ISO COOLING	207,217	0	2,921	0	0	210,138
RESIDUAL HEAT REMOVAL	1,112,292	0	13,935	0	0	1,126,227
POOL LINER & RACKS	6,599,226	0	85,614	0	0	6,684,841
CONTAMINATED CONCRETE	7,497,463	0	97,526	0	0	7,594,988
OTHER REACTOR BUILDING	19,329,718	0	318,850	0	0	19,648,567
TURBINE	25,149,548	0	315,936	0	0	25,465,483
NUCLEAR STEAM CONDENSATE	5,444,142	0	81,567	0	0	5,525,710
LOW PRESSURE FEEDWATER HEATERS	12,506,591	0	165,603	0	0	12,672,194
MAIN STEAM	1,009,776	0	15,958	0	0	1,025,735
MOISTURE SEPARATOR REHEATERS	10,337,603	0	160,666	0	0	10,498,269
REACTOR FEEDWATER PUMPS	2,719,222	0	43,593	0	0	2,762,815
HIGH PRESSURE FEEDWATER HEATERS	1,992,572	0	27,189	0	0	2,019,761
OTHER TG BLDG	74,855,064	0	1,091,382	0	0	75,946,446
RAD WASTE BLDG	34,088,712	0	540,416	0	0	34,629,128
REACTOR BLDG	9,219,561	4,525,120	85,500	0	0	13,830,181
TG BLDG	6,065,500	2,969,610	56,250	0	0	9,091,360
RAD WASTE & CONTROL	5,580,260	2,686,790	51,750	0	0	8,318,800
CONCENTRATOR BOTTOMS	20,621,513	15,908,625	40,690,470	0	2,001,662	79,222,269
OTHER	5,590,721	4,313,005	431,550	0	0	10,335,276
POST-TMI-2 ADDITIONS	470,360	0	0	0	0	470,360
SUBTOTAL BWR COSTS	**330,008,637**	**56,351,885**	**105,702,201**	**0**	**3,032,410**	**495,095,133**
ATLANTIC COMPACT COMMISSION ADMINISTRATIVE SURCHARGE						4,021,086
TOTAL BWR COSTS (OUTSIDE COMPACT)						**499,116,219**

(a) GTCC Material: Assumes a low-density, distr buted packaging scheme and final disposal as LLW. High-density packaging, ISFSI storage, and geologic repository disposal could reduce disposal costs.

Table B-23. PWR Burial Costs at the South Carolina Site Atlantic Compact (2004 dollars)

REFERENCE PWR COMPONENT	BASE DISPOSAL CHARGE	CASK HANDLING	CURIE SURCHARGE	LINER DOSE RATE	DOSE RATE SURCHARGE	DISPOSAL COST
VESSEL WALL	2,838,980	2,061,272	5,441,752	0	1,362,711	11,704,715
VESSEL HEAD & BOTTOM	1,808,550	2,169,760	7,160	0	0	3,985,470
UPPER CORE SUPPORT ASSM	170,740	216,976	3,580	0	54,637	445,932
UPPER SUPPORT COLUMN	157,854	216,976	35,800	0	50,513	461,143
UPPER CORE BARREL	75,177	108,488	286,408	0	36,085	506,158
UPPER CORE GRID PLATE	187,943	271,220	716,020	0	90,212	1,265,395
GUIDE TUBES	278,155	325,464	35,800	0	75,102	714,521
LOWER CORE BARREL[a]	1,202,832	1,735,808	4,582,528	0	577,359	8,098,527
THERMAL SHIELDS[a]	225,531	325,464	859,224	0	108,255	1,518,474
CORE SHROUD[a]	174,605	216,976	8,735,444	0	83,811	9,210,836
LOWER GRID PLATE[a]	187,943	271,220	1,432,040	0	90,212	1,981,415
LOWER SUPPORT COLUMN	47,678	54,244	143,204	0	22,886	268,012
LOWER CORE FORGING	518,017	596,684	895,000	0	248,648	2,258,349
MISC INTERNALS	420,000	433,952	716,000	0	201,600	1,771,552
BIO SHIELD CONCRETE	10,237,500	0	429,600	0	0	10,667,100
REACTOR CAVITY LINER	206,176	0	3,580	0	0	209,756
REACTOR COOLANT PUMPS	3,589,296	0	27,802	0	0	3,617,098
PRESSURIZER	1,628,835	0	1,815	0	0	1,630,650
R.Hx, EHx, SUMP PUMP, CAVITY PUMP	150,354	0	4,221	0	0	154,575
PRESSURIZER RELIEF TANK	451,062	0	1,446	0	0	452,508
SAFETY INJECTION ACCUM TANKS	1,606,500	0	29,156	0	0	1,635,656
STEAM GENERATORS	13,135,296	0	1,575,200	0	0	14,710,496
REACTOR COOLANT PIPING	1,430,333	0	106,684	0	0	1,537,017
REMAINING CONTAM. MATLS	25,285,554	0	80,117	0	0	25,365,670
CONTAM. MATL OTHER BLDG	194,522,610	0	65,962	0	0	194,588,572
FILTER CARTRIDGES	257,742	325,464	1,790,000	0	30,929	2,404,135
SPENT RESINS	945,000	1,084,880	2,864,080	0	453,600	5,347,560
COMBUSTIBLE WASTES	4,510,620	3,254,640	107,400	0	0	7,872,660
EVAPORATOR BOTTOMS	4,441,500	5,098,936	13,461,176	0	606,690	23,608,302
POST-TMI-2 ADDITIONS	8,913,864	0	0	0	0	8,913,864
SUBTOTAL PWR COSTS	**279,606,246**	**18,768,424**	**44,438,198**	**0**	**4,093,250**	**346,906,118**
ATLANTIC COMPACT SURCHARGE						3,883,482
TOTAL PWR COSTS (INSIDE COMPACT)						**350,789,600**

(a) GTCC Material: Assumes a low-density, distributed packaging scheme and final disposal as LLW. High-density packaging, ISFSI storage, and geologic repository disposal could reduce disposal costs.

NUREG-1307

Table B-24. BWR Burial Costs at the South Carolina Site Atlantic Compact (2004 dollars)

REFERENCE BWR COMPONENT	BASE DISPOSAL CHARGE	CASK HANDLING	CURIE SURCHARGE	LINER DOSE RATE	DOSE RATE SURCHARGE	DISPOSAL COST
STEAM SEPARATOR	174,477	1,518,832	2,004,856	0	83,749	3,781,915
FUEL SUPPORT & PIECES	76,852	759,416	250,600	0	36,889	1,123,757
CONTROL RODS/INCORES	228,816	433,952	1,145,632	0	109,832	1,918,232
CONTROL RODS GUIDES	64,318	650,928	35,800	0	23,798	774,844
JET PUMPS	186,063	2,169,760	2,864,080	0	89,310	5,309,213
TOP FUEL GUIDES	318,750	3,905,568	10,310,688	0	153,000	14,688,007
CORE SUPPORT PLATE	213,675	1,681,564	232,700	0	79,060	2,206,999
CORE SHROUD[(a)]	623,969	7,594,160	20,048,560	0	299,505	28,566,194
REACTOR VESSEL WALL	135,741	1,193,368	773,280	0	50,224	2,152,614
SAC SHIELD (NEUTRON ACTIV. MATL)	2,856,173	0	60,144	0	0	2,916,317
REACTOR WATER REC	1,235,742	0	15,733	0	0	1,251,475
SAC SHIELD (CONTAM. MATL)	7,396,623	0	55,420	0	0	7,452,043
OTHER PRIMARY CONTAINMENT	51,877,142	0	632,107	0	0	52,509,249
CONTAINMENT ATMOSPHERIC	637,125	0	8,580	0	0	645,705
HIGH PRESSURE CORE SPRAY	320,250	0	3,037	0	0	323,287
LOW PRESSURE CORE SPRAY	141,746	0	1,787	0	0	143,533
REACTOR BLDG CLOSED COOLING	499,844	0	5,720	0	0	505,564
REACTOR CORE ISO COOLING	172,531	0	2,324	0	0	174,855
RESIDUAL HEAT REMOVAL	966,011	0	11,086	0	0	977,097
POOL LINER & RACKS	6,371,167	0	68,111	0	0	6,439,278
CONTAMINATED CONCRETE	7,026,349	0	77,587	0	0	7,103,936
OTHER REACTOR BUILDING	18,835,597	0	253,663	0	0	19,089,260
TURBINE	23,569,241	0	251,344	0	0	23,820,586
NUCLEAR STEAM CONDENSATE	4,818,470	0	64,891	0	0	4,883,361
LOW PRESSURE FEEDWATER HEATERS	10,460,855	0	131,747	0	0	10,592,602
MAIN STEAM	942,720	0	12,696	0	0	955,415
MOISTURE SEPARATOR REHEATERS	9,491,096	0	127,819	0	0	9,618,915
REACTOR FEEDWATER PUMPS	2,575,188	0	34,681	0	0	2,609,869
HIGH PRESSURE FEEDWATER HEATERS	1,697,388	0	21,630	0	0	1,719,018
OTHER TG BLDG	64,471,795	0	868,255	0	0	65,340,050
RAD WASTE BLDG	31,924,289	0	429,931	0	0	32,354,220
REACTOR BLDG	8,161,158	3,471,616	68,020	0	0	11,700,794
TG BLDG	5,369,183	2,278,248	44,750	0	0	7,692,181
RAD WASTE & CONTROL	4,939,648	2,061,272	41,170	0	0	7,042,090
CONCENTRATOR BOTTOMS	18,254,169	12,204,900	32,220,900	0	2,472,831	65,152,801
OTHER	4,948,908	3,308,884	343,322	0	123,317	8,724,431
POST-TMI-2 ADDITIONS	728,551	0	0	0	0	728,551
SUBTOTAL BWR COSTS	**292,711,621**	**43,232,468**	**73,522,651**	**0**	**3,521,516**	**412,988,255**
ATLANTIC COMPACT SURCHARGE						4,021,086
TOTAL BWR COSTS (INSIDE COMPACT)						**417,009,341**

(a) GTCC Material: Assumes a low-density, distr buted packaging scheme and final disposal as LLW. High-density packaging, ISFSI storage, and geologic repository disposal could reduce disposal costs.

Table B-25. PWR Burial Costs at the South Carolina Site Non-Atlantic Compact (2004 dollars)

REFERENCE PWR COMPONENT	BASE DISPOSAL CHARGE	CASK HANDLING	CURIE SURCHARGE	LINER DOSE RATE	DOSE RATE SURCHARGE	DISPOSAL COST
VESSEL WALL	2,841,954	2,380,320	6,064,800	0	1,364,138	12,651,213
VESSEL HEAD & BOTTOM	1,709,463	2,505,600	7,980	0	0	4,223,043
UPPER CORE SUPPORT ASSM	169,733	250,560	3,990	0	54,314	478,597
UPPER SUPPORT COLUMN	167,213	250,560	39,900	0	53,508	511,181
UPPER CORE BARREL	63,000	125,280	319,200	0	30,240	537,720
UPPER CORE GRID PLATE	144,585	313,200	798,000	0	69,401	1,325,186
GUIDE TUBES	255,245	375,840	39,900	0	68,916	739,901
LOWER CORE BARREL[a]	1,184,400	2,004,480	5,107,200	0	568,512	8,864,592
THERMAL SHIELDS[a]	229,425	375,840	957,600	0	110,124	1,672,989
CORE SHROUD[a]	173,576	250,560	9,735,600	0	83,316	10,243,052
LOWER GRID PLATE[a]	164,430	313,200	1,596,000	0	78,926	2,152,556
LOWER SUPPORT COLUMN	45,066	62,640	159,600	0	21,632	288,938
LOWER CORE FORGING	489,636	689,040	997,500	0	235,025	2,411,201
MISC. INTERNALS	403,200	501,120	798,000	0	193,536	1,895,856
BIO SHIELD CONCRETE	10,647,000	0	478,800	0	0	11,125,800
REACTOR CAVITY LINER	218,400	0	3,990	0	0	222,390
REACTOR COOLANT PUMPS	3,593,056	0	30,986	0	0	3,624,042
PRESSURIZER	1,638,000	0	2,023	0	0	1,640,023
R.Hx, EHx, SUMP PUMP, CAVITY PUMP	148,680	0	4,704	0	0	153,384
PRESSURIZER RELIEF TANK	385,560	0	1,612	0	0	387,172
SAFETY INJECTION ACCUM. TANKS	1,670,760	0	32,495	0	0	1,703,255
STEAM GENERATORS	13,438,016	0	1,755,600	0	0	15,193,616
REACTOR COOLANT PIPING	1,421,897	0	118,902	0	0	1,540,799
REMAINING CONTAM. MATLS	23,900,205	0	89,292	0	0	23,989,497
CONTAM. MATRL OTHER BLDG	206,055,691	0	0	0	0	206,055,691
FILTER CARTRIDGES	258,012	375,840	1,995,000	0	20,641	2,649,493
SPENT RESINS	935,640	1,252,800	3,192,000	0	449,107	5,829,547
COMBUSTIBLE WASTES	4,536,000	3,758,400	119,700	0	0	8,414,100
EVAPORATOR BOTTOMS	4,397,508	5,888,160	15,002,400	0	432,266	25,720,334
POST-TMI-2 ADDITIONS	5,098,439	0	0	0	0	5,098,439
SUBTOTAL PWR COSTS	286,383,788	21,673,440	49,452,774	0	3,833,603	361,343,605
ATLANTIC COMPACT SURCHARGE						3,883,482
TOTAL PWR COSTS (OUTSIDE COMPACT)						365,227,087

(a) GTCC Material: Assumes a low-density, distributed packaging scheme and final disposal as LLW. High-density packaging, ISFSI storage, and geologic repository disposal could reduce disposal costs.

Disposal Cost Based on Flat Rate Calculation

Base Cost = (Waste Volume [ft^3]) * \$600/ft^3 = 645,247 * 600 =   387,148,200
Spent Resins = (Resin Volume [ft^3]) * \$1,800/ft^3 = 2000 * 1,800 =   3,600,000
Atlantic Compact Surcharge = Volume [ft^3] * \$6 ft^3 = 647,247 * 6 =   3,883,482
Total   **394,631,682**

Table B-26. BWR Burial Costs at the South Carolina Site Non-Atlantic Compact (2004 dollars)

REFERENCE BWR COMPONENT	BASE DISPOSAL CHARGE	CASK HANDLING	CURIE SURCHARGE	LINER DOSE RATE	DOSE RATE SURCHARGE	DISPOSAL COST
STEAM SEPARATOR	160,107	1,753,920	2,234,400	0	76,851	4,225,278
FUEL SUPPORT & PIECES	76,399	876,960	279,300	0	36,671	1,269,330
CONTROL RODS/INCORES	233,392	501,120	1,276,800	0	112,028	2,123,341
CONTROL RODS GUIDES	64,680	751,680	39,900	0	23,932	880,192
JET PUMPS	166,320	2,505,600	3,192,000	0	79,834	5,943,754
TOP FUEL GUIDES	293,832	4,510,080	11,491,200	0	141,039	16,436,151
CORE SUPPORT PLATE	205,128	1,941,840	259,350	0	75,897	2,482,215
CORE SHROUD[(a)]	665,469	8,769,600	22,344,000	0	319,425	32,098,494
REACTOR VESSEL WALL	128,304	1,378,080	861,840	0	47,473	2,415,697
SAC SHIELD (NEUTRON ACTIV. MATL.)	2,859,165	0	67,032	0	0	2,926,197
REACTOR WATER REC	1,309,008	0	17,535	0	0	1,326,543
SAC SHIELD (CONTAM. MATL)	7,404,371	0	61,767	0	0	7,466,138
OTHER PRIMARY CONTAINMENT	55,429,605	0	704,499	0	0	56,134,104
CONTAINMENT ATMOSPHERIC	577,500	0	9,563	0	0	587,063
HIGH PRESSURE CORE SPRAY	307,440	0	3,385	0	0	310,825
LOW PRESSURE CORE SPRAY	150,150	0	1,992	0	0	152,142
REACTOR BLDG CLOSED COOLING	502,656	0	6,375	0	0	509,031
REACTOR CORE ISO COOLING	183,576	0	2,590	0	0	186,165
RESIDUAL HEAT REMOVAL	985,331	0	12,356	0	0	997,686
POOL LINER & RACKS	5,846,402	0	75,911	0	0	5,922,313
CONTAMINATED CONCRETE	6,641,389	0	86,473	0	0	6,727,861
OTHER REACTOR BUILDING	17,125,189	0	282,714	0	0	17,407,903
TURBINE	22,277,926	0	280,130	0	0	22,558,056
NUCLEAR STEAM CONDENSATE	4,822,853	0	72,323	0	0	4,895,176
LOW PRESSURE FEEDWATER HEATERS	11,081,070	0	146,835	0	0	11,227,905
MAIN STEAM	894,613	0	14,150	0	0	908,762
MOISTURE SEPARATOR REHEATERS	9,157,868	0	142,457	0	0	9,300,325
REACTOR FEEDWATER PUMPS	2,408,903	0	38,652	0	0	2,447,555
HIGH PRESSURE FEEDWATER HEATERS	1,765,284	0	24,108	0	0	1,789,391
OTHER TG BLDG	66,314,836	0	967,692	0	0	67,282,528
RAD WASTE BLDG	30,200,940	0	479,169	0	0	30,680,109
REACTOR BLDG	8,169,707	4,008,960	75,810	0	0	12,254,477
TG BLDG	5,374,807	2,630,880	49,875	0	0	8,055,562
RAD WASTE & CONTROL	4,944,823	2,380,320	45,885	0	0	7,371,028
CONCENTRATOR BOTTOMS	18,273,292	14,094,000	35,910,000	0	1,773,728	70,051,019
OTHER	4,954,092	3,821,040	382,641	0	0	9,157,773
POST-TMI-2 ADDITIONS	416,707	0	0	0	0	416,707
SUBTOTAL BWR COSTS	**292,373,132**	**49,924,080**	**81,940,707**	**0**	**2,686,878**	**426,924,797**
ATLANTIC COMPACT SURCHARGE						4,021,086
TOTAL BWR COSTS (OUTSIDE COMPACT)						**430,945,883**

(a) GTCC Material: Assumes a low-density, distributed packaging scheme and final disposal as LLW. High-density packaging, ISFSI storage, and geologic repository disposal could reduce disposal costs.

Disposal Cost Based on Flat Rate Calculation

Base Cost = (Waste Volume [ft^3]) * \$600/ft^3 = 670,181 * 600 =   402,108,600

Spent Resins = (Resin Volume [ft^3]) * \$1,800/ft^3 = 0 * 1,800 =   0

Atlantic Compact Surcharge = Volume [ft^3] * \$6 ft^3 = 670,181 * 6 =   4,021,086

Total   **406,129,686**

Table B-27. PWR Burial Costs at the South Carolina Site Atlantic Compact (2002 dollars)

REFERENCE PWR COMPONENT	BASE DISPOSAL CHARGE	CASK HANDLING	CURIE SURCHARGE	LINER DOSE RATE	DOSE RATE SURCHARGE	DISPOSAL COST
VESSEL WALL	2,617,120	1,900,304	5,016,760	0	1,256,218	10,790,402
VESSEL HEAD & BOTTOM	1,667,358	2,000,320	6,600	0	0	3,674,278
UPPER CORE SUPPORT ASSM	157,410	200,032	3,300	0	50,371	411,113
UPPER SUPPORT COLUMN	145,530	200,032	33,000	0	46,570	425,132
UPPER CORE BARREL	69,300	100,016	264,040	0	33,264	466,620
UPPER CORE GRID PLATE	173,250	250,040	660,100	0	83,160	1,166,550
GUIDE TUBES	256,410	300,048	33,000	0	69,231	658,689
LOWER CORE BARREL[a]	1,108,800	1,600,256	4,224,640	0	532,224	7,465,920
THERMAL SHIELDS[a]	207,900	300,048	792,120	0	99,792	1,399,860
CORE SHROUD[a]	160,974	200,032	8,053,220	0	77,268	8,491,494
LOWER GRID PLATE[a]	173,250	250,040	1,320,200	0	83,160	1,826,650
LOWER SUPPORT COLUMN	43,956	50,008	132,020	0	21,099	247,083
LOWER CORE FORGING	477,576	550,088	825,000	0	229,236	2,081,900
MISC. INTERNALS	387,200	400,064	660,000	0	185,856	1,633,120
BIO SHIELD CONCRETE	9,438,000	0	396,000	0	0	9,834,000
REACTOR CAVITY LINER	190,080	0	3,300	0	0	193,380
REACTOR COOLANT PUMPS	3,308,800	0	25,628	0	0	3,334,428
PRESSURIZER	1,501,500	0	1,673	0	0	1,503,173
R.Hx, EHx, SUMP PUMP, CAVITY PUMP	138,600	0	3,891	0	0	142,491
PRESSURIZER RELIEF TANK	415,800	0	1,333	0	0	417,133
SAFETY INJECTION ACCUM TANKS	1,481,040	0	26,875	0	0	1,507,915
STEAM GENERATORS	12,108,800	0	1,452,000	0	0	13,560,800
REACTOR COOLANT PIPING	1,318,668	0	98,340	0	0	1,417,008
REMAINING CONTAM. MATLS	23,311,530	0	73,851	0	0	23,385,381
CONTAM. MATL OTHER BLDG	179,336,381	0	60,803	0	0	179,397,184
FILTER CARTRIDGES	237,600	300,048	1,650,000	0	28,512	2,216,160
SPENT RESINS	871,200	1,000,160	2,640,400	0	418,176	4,929,936
COMBUSTIBLE WASTES	4,158,000	3,000,480	99,000	0	0	7,257,480
EVAPORATOR BOTTOMS	4,094,640	4,700,752	12,409,880	0	559,310	21,764,582
POST-TMI-2 ADDITIONS	8,217,949	0	0	0	0	8,217,949
SUBTOTAL PWR COSTS	**257,774,622**	**17,302,768**	**40,966,973**	**0**	**3,773,446**	**319,817,810**
ATLANTIC COMPACT SURCHARGE (INSIDE COMPACT)						2,588,988
TOTAL PWR COSTS (INSIDE COMPACT)						322,406,798

(a) GTCC Material: Assumes a low-density, distributed packaging scheme and final disposal as LLW. High-density packaging, ISFSI storage, and geologic repository disposal could reduce disposal costs.

Table B-28. BWR Burial Costs at the South Carolina Site Atlantic Compact (2002 dollars)

REFERENCE BWR COMPONENT	BASE DISPOSAL CHARGE	CASK HANDLING	CURIE SURCHARGE	LINER DOSE RATE	DOSE RATE SURCHARGE	DISPOSAL COST
STEAM SEPARATOR	160,838	1,400,224	1,848,280	0	77,202	3,486,544
FUEL SUPPORT & PIECES	70,852	700,112	231,000	0	34,009	1,035,973
CONTROL RODS/INCORES	210,947	400,064	1,056,160	0	101,254	1,768,425
CONTROL RODS GUIDES	59,290	600,096	33,000	0	21,937	714,323
JET PUMPS	171,518	2,000,320	2,640,400	0	82,328	4,894,566
TOP FUEL GUIDES	293,832	3,600,576	9,505,440	0	141,039	13,540,887
CORE SUPPORT PLATE	196,988	1,550,248	214,500	0	72,886	2,034,622
CORE SHROUD[(a)]	575,190	7,001,120	18,482,800	0	276,091	26,335,201
REACTOR VESSEL WALL	125,144	1,100,176	712,800	0	46,303	1,984,423
SAC SHIELD (NEUTRON ACTIV. MATL.)	2,632,969	0	55,440	0	0	2,688,409
REACT. WATER REC	1,139,268	0	14,503	0	0	1,153,771
SAC SHIELD (CONTAM. MATL.)	6,818,592	0	51,086	0	0	6,869,678
OTHER PRIMARY CONTAINMENT	47,821,620	0	582,668	0	0	48,404,288
CONTAINM. ATMOSPHERIC	587,318	0	7,909	0	0	595,227
HIGH PRESSURE CORE SPRAY	295,240	0	2,800	0	0	298,040
LOW PRESSURE CORE SPRAY	130,680	0	1,647	0	0	132,327
REACTOR BLDG CLOSED COOLING	460,768	0	5,273	0	0	466,041
REACTOR CORE ISO COOLING	159,044	0	2,142	0	0	161,185
RESIDUAL HEAT REMOVAL	890,570	0	10,219	0	0	900,789
POOL LINER & RACKS	5,873,098	0	62,784	0	0	5,935,882
CONTAMINATED CONCRETE	6,477,808	0	71,519	0	0	6,549,326
OTHER REACTOR BUILDING	17,363,115	0	233,823	0	0	17,596,938
TURBINE	21,729,209	0	231,686	0	0	21,960,895
NUCLEAR STEAM CONDENSATE	4,441,784	0	59,816	0	0	4,501,600
LOW PRESSURE FEEDWATER HEATERS	9,644,184	0	121,443	0	0	9,765,627
MAIN STEAM	869,022	0	11,703	0	0	880,725
MOISTURE SEPARATOR REHEATERS	8,749,125	0	117,822	0	0	8,866,947
REACTOR FEEDWATER PUMPS	2,373,872	0	31,968	0	0	2,405,840
HIGH PRESSURE FEEDWATER HEATERS	1,564,830	0	19,939	0	0	1,584,769
OTHER TG BLDG	59,431,680	0	800,347	0	0	60,232,027
RAD WASTE BLDG	29,428,592	0	396,305	0	0	29,824,897
REACTOR BLDG	7,523,380	3,200,512	62,700	0	0	10,786,592
TG BLDG	4,949,592	2,100,336	41,250	0	0	7,091,178
RAD WASTE & CONTROL	4,553,625	1,900,304	37,950	0	0	6,491,879
CONCENTRATOR BOTTOMS	16,827,644	11,251,800	29,704,500	0	2,279,585	60,063,529
OTHER	4,562,161	3,050,488	316,470	0	113,680	8,042,799
POST-TMI-2 ADDITIONS	671,672	0	0	0	0	671,672
SUBTOTAL BWR COSTS	269,835,058	39,856,376	67,780,090	0	3,246,316	380,717,839
ATLANTIC COMPACT SURCHARGE (INSIDE COMPACT)						2,680,724
TOTAL BWR COSTS (INSIDE COMPACT)						**383,398,563**

(a) GTCC Material: Assumes a low-density, distributed packaging scheme and final disposal as LLW. High-density packaging, ISFSI storage, and geologic repository disposal could reduce disposal costs.

Table B-29. BWR Burial Costs at the South Carolina Site Non-Atlantic Compact (2002 dollars)

REFERENCE PWR COMPONENT	BASE DISPOSAL CHARGE	CASK HANDLING	CURIE SURCHARGE	LINER DOSE RATE	DOSE RATE SURCHARGE	DISPOSAL COST
VESSEL WALL	2,730,132	1,983,600	5,236,704	0	1,310,463	11,260,899
VESSEL HEAD & BOTTOM	1,740,340	2,088,000	7,600	0	0	3,835,940
UPPER CORE SUPPORT ASSM	164,300	208,800	3,800	0	52,576	429,476
UPPER SUPPORT COLUMN	151,900	208,800	38,000	0	48,608	447,308
UPPER CORE BARREL	72,360	104,400	275,616	0	34,733	487,109
UPPER CORE GRID PLATE	180,900	261,000	689,040	0	86,832	1,217,772
GUIDE TUBES	267,732	313,200	38,000	0	72,288	691,220
LOWER CORE BARREL[a]	1,157,760	1,670,400	4,409,856	0	555,725	7,793,741
THERMAL SHIELDS[a]	217,080	313,200	826,848	0	104,198	1,461,326
CORE SHROUD[a]	168,020	208,800	8,406,288	0	80,650	8,863,758
LOWER GRID PLATE[a]	180,900	261,000	1,378,080	0	86,832	1,906,812
LOWER SUPPORT COLUMN	45,880	52,200	137,808	0	22,022	257,910
LOWER CORE FORGING	498,480	574,200	950,000	0	239,270	2,261,950
MISC. INTERNALS	404,000	417,600	760,000	0	193,920	1,775,520
BIO SHIELD CONCRETE	9,847,500	0	456,000	0	0	10,303,500
REACTOR CAVITY LINER	198,400	0	3,800	0	0	202,200
REACTOR COOLANT PUMPS	3,451,680	0	29,511	0	0	3,481,191
PRESSURIZER	1,567,800	0	1,927	0	0	1,569,727
R.Hx, EHx, SUMP PUMP, CAVITY PUMP	144,720	0	4,480	0	0	149,200
PRESSURIZER RELIEF TANK	434,160	0	1,535	0	0	435,695
SAFETY INJECTION ACCUM. TANKS	1,545,300	0	30,947	0	0	1,576,247
STEAM GENERATORS	12,631,680	0	1,672,000	0	0	14,303,680
REACTOR COOLANT PIPING	1,376,388	0	113,240	0	0	1,489,628
REMAINING CONTAM. MATLS	24,331,900	0	85,040	0	0	24,416,940
CONTAM. MATL OTHER BLDG	187,186,122	0	70,015	0	0	187,256,137
FILTER CARTRIDGES	247,860	313,200	1,900,000	0	29,743	2,490,803
SPENT RESINS	909,000	1,044,000	2,756,160	0	436,320	5,145,480
COMBUSTIBLE WASTES	4,341,600	3,132,000	114,000	0	0	7,587,600
EVAPORATOR BOTTOMS	4,272,300	4,906,800	12,953,952	0	583,578	22,716,630
POST-TMI-2 ADDITIONS	8,572,815	0	0	0	0	8,572,815
SUBTOTAL PWR COSTS	**269,039,008**	**18,061,200**	**43,350,247**	**0**	**3,937,759**	**334,388,214**
ATLANTIC COMPACT SURCHARGE (OUTSIDE COMPACT)						2,588,988
TOTAL PWR COSTS (OUTSIDE COMPACT)						**336,977,202**

(a) GTCC Material: Assumes a low-density, distributed packaging scheme and final disposal as LLW. High-density packaging, ISFSI storage, and geologic repository disposal could reduce disposal costs.

Table B-30. BWR Burial Costs at the South Carolina Site Non-Atlantic Compact (2002 dollars)

REFERENCE BWR COMPONENT	BASE DISPOSAL CHARGE	CASK HANDLING	CURIE SURCHARGE	LINER DOSE RATE	DOSE RATE SURCHARGE	DISPOSAL COST
STEAM SEPARATOR	167,940	1,461,600	1,929,312	0	80,611	3,639,462
FUEL SUPPORT & PIECES	73,954	730,800	266,000	0	35,498	1,106,251
CONTROL RODS/INCORES	220,099	417,600	1,102,464	0	105,648	1,845,811
CONTROL RODS GUIDES	61,908	626,400	38,000	0	22,906	749,214
JET PUMPS	179,091	2,088,000	2,756,160	0	85,964	5,109,215
TOP FUEL GUIDES	306,806	3,758,400	9,922,176	0	147,267	14,134,649
CORE SUPPORT PLATE	205,535	1,618,200	247,000	0	76,048	2,146,783
CORE SHROUD[(a)]	600,588	7,308,000	19,293,120	0	288,282	27,489,990
REACTOR VESSEL WALL	130,622	1,148,400	820,800	0	48,330	2,148,152
SAC SHIELD (NEUTRON ACTIV. MATL.)	2,746,665	0	63,840	0	0	2,810,505
REACT. WATER REC	1,189,135	0	16,700	0	0	1,205,835
SAC SHIELD (CONTAM. MATL.)	7,113,031	0	58,826	0	0	7,171,857
OTHER PRIMARY CONTAINMENT	49,933,224	0	670,951	0	0	50,604,175
CONTAINM. ATMOSPHERIC	613,251	0	9,108	0	0	622,359
HIGH PRESSURE CORE SPRAY	308,050	0	3,224	0	0	311,274
LOW PRESSURE CORE SPRAY	136,400	0	1,897	0	0	138,297
REACTOR BLDG CLOSED COOLING	481,114	0	6,072	0	0	487,185
REACTOR CORE ISO COOLING	166,066	0	2,466	0	0	168,532
RESIDUAL HEAT REMOVAL	929,210	0	11,767	0	0	940,977
POOL LINER & RACKS	6,132,430	0	72,296	0	0	6,204,726
CONTAMINATED CONCRETE	6,761,348	0	82,355	0	0	6,843,703
OTHER REACTOR BUILDING	18,129,798	0	269,251	0	0	18,399,049
TURBINE	22,680,319	0	266,790	0	0	22,947,109
NUCLEAR STEAM CONDENSATE	4,637,914	0	68,879	0	0	4,706,793
LOW PRESSURE FEEDWATER HEATERS	10,066,320	0	139,843	0	0	10,206,163
MAIN STEAM	907,394	0	13,476	0	0	920,870
MOISTURE SEPARATOR REHEATERS	9,135,450	0	135,673	0	0	9,271,123
REACTOR FEEDWATER PUMPS	2,478,692	0	36,812	0	0	2,515,504
HIGH PRESSURE FEEDWATER HEATERS	1,632,726	0	22,960	0	0	1,655,685
OTHER TG BLDG	62,055,936	0	921,611	0	0	62,977,547
RAD WASTE BLDG	30,728,036	0	456,351	0	0	31,184,387
REACTOR BLDG	7,848,254	3,340,800	72,200	0	0	11,261,254
TG BLDG	5,163,325	2,192,400	47,500	0	0	7,403,225
RAD WASTE & CONTROL	4,750,259	1,983,600	43,700	0	0	6,777,559
CONCENTRATOR BOTTOMS	17,554,292	11,745,000	31,006,800	0	2,378,021	62,684,114
OTHER	4,759,164	3,184,200	364,420	0	118,589	8,426,373
POST-TMI-2 ADDITIONS	700,676	0	0	0	0	700,676
SUBTOTAL BWR COSTS	**281,685,021**	**41,603,400**	**71,240,801**	**0**	**3,387,164**	**397,916,385**
ATLANTIC COMPACT SURCHARGE (OUTSIDE COMPACT)						2,680,724
TOTAL BWR COSTS (OUTSIDE COMPACT)						**400,597,109**

(a) GTCC Material: Assumes a low-density, distributed packaging scheme and final disposal as LLW. High-density packaging, ISFSI storage, and geologic repository disposal could reduce disposal costs.

NUREG-1307

Table B-31. PWR LLW Disposition Costs Using a Combination of Non-Compact Disposal Facility and the Washington Disposal Facility (2012 dollars)

REFERENCE PWR COMPONENT	VOLUME CHARGE	SHIPMENT CHARGE	CONTAINER CHARGE	CONTAINER DOSE RATE CHARGE	NON-COMPACT CHARGE	DISPOSAL COST
VESSEL WALL	351,120	522,500	287,280	777,649	0	1,938,549
VESSEL HEAD & BOTTOM	369,600	550,000	302,400	0	0	1,222,000
UPPER CORE SUPPORT ASSM	36,960	55,000	30,240	0	0	122,200
UPPER SUPPORT COLUMN	36,960	55,000	30,240	0	0	122,200
UPPER CORE BARREL	18,480	27,500	15,120	0	0	61,100
UPPER CORE GRID PLATE	46,200	68,750	37,800	0	0	152,750
GUIDE TUBES	55,440	82,500	45,360	0	0	183,300
LOWER CORE BARREL[a]	295,680	440,000	241,920	0	0	977,600
THERMAL SHIELDS[a]	55,440	82,500	45,360	0	0	183,300
CORE SHROUD[a]	36,960	55,000	30,240	0	0	122,200
LOWER GRID PLATE[a]	46,200	68,750	37,800	0	0	152,750
LOWER SUPPORT COLUMN	9,240	13,750	7,560	0	0	30,550
LOWER CORE FORGING	101,640	151,250	83,160	0	0	336,050
MISC INTERNALS	73,920	110,000	60,480	0	0	244,400
BIO SHIELD CONCRETE	0	0	0	0	4,530,240	4,530,240
REACTOR CAVITY LINER	47,309	13,750	30,240	0	0	91,299
REACTOR COOLANT PUMPS	0	0	0	0	1,617,000	1,617,000
PRESSURIZER	0	0	0	0	1,386,000	1,386,000
R.Hx,EHx,SUMP PUMP, CAVITY PUMP	0	0	0	0	72,600	72,600
PRESSURIZER RELIEF TANK	0	0	0	0	217,800	217,800
SAFETY INJECTION ACCUM TANKS	0	0	0	0	726,000	726,000
STEAM GENERATORS	0	0	0	0	8,224,370	8,224,370
REACTOR COOLANT PIPING	0	0	0	0	598,950	598,950
REMAINING CONTAM. MATLS	0	0	0	0	9,548,352	9,548,352
CONTAMINATED MATRL OTHR BLD	0	0	0	0	86,595,647	86,595,647
FILTER CARTRIDGES	0	0	0	0	159,390	159,390
SPENT RESINS	184,800	275,000	151,200	0	0	611,000
COMBUSTIBLE WASTES	0	0	0	0	6,404,063	6,404,063
EVAPORATOR BOTTOMS	868,560	1,292,500	710,640	0	0	2,871,700
POST-TMI-2 ADDITIONS	1,438,021	0	0	0	0	1,438,021
HEAVY OBJECT SURCHARGE						0
SITE AVAILABILITY CHARGES						399,078
SUBTOTAL PWR COSTS	**4,072,530**	**3,863,750**	**2,147,040**	**777,649**	**120,080,411**	**131,340,458**
TAXES & FEES (% OF CHARGES)						484,182
TAXES & FEES ($/UNIT VOL.)						725,794
ANNUAL PERMIT FEES (3 YRS)						127,200
TOTAL PWR COSTS						**132,677,634**

(a) GTCC Material: Assumes a low-density, distr buted packaging scheme and final disposal as LLW. High-density packaging, ISFSI storage, and geologic repository disposal could reduce disposal costs.

Table B-32. BWR LLW Disposition Costs Using a Combination of Non-Compact Disposal Facility and the Washington Disposal Facility (2012 dollars)

REFERENCE BWR COMPONENT	VOLUME CHARGE	SHIPMENT CHARGE	CONTAINER CHARGE	CONTAINER DOSE RATE CHARGE	NON-COMPACT CHARGE	DISPOSAL COST
STEAM SEPARATOR	32,631	192,500	211,680	777,649	0	1,214,460
FUEL SUPPORT & PIECES	16,315	96,250	105,840	0	0	218,405
CONTROL RODS/INCORES	48,946	110,000	60,480	0	0	219,426
CONTROL RODS GUIDES	13,052	82,500	90,720	0	0	186,272
JET PUMPS	45,683	275,000	302,400	0	0	623,083
TOP FUEL GUIDES	78,314	990,000	544,320	0	0	1,612,634
CORE SUPPORT PLATE	35,894	220,000	234,360	0	0	490,254
CORE SHROUD[a]	153,364	1,925,000	1,058,400	0	0	3,136,764
REACTOR VESSEL WALL	26,105	275,000	166,320	0	0	467,425
SAC SHIELD Neutron-Activated Matl	0	0	0	0	576,864	576,864
REACT. WATER REC	0	0	0	0	564,045	564,045
SAC SHIELD Contaminated Matl	0	0	0	0	1,986,976	1,986,976
OTHER PRIMARY CONTAINMENT	0	0	0	0	22,664,345	22,664,345
CONTAINM. ATMOSPHERIC	0	0	0	0	307,661	307,661
HIGH PRESSURE CORE SPRAY	0	0	0	0	108,963	108,963
LOW PRESSURE CORE SPRAY	0	0	0	0	64,096	64,096
REACTOR BLDG CLOSED COOLING	0	0	0	0	205,107	205,107
REACTOR CORE ISO COOLING	0	0	0	0	83,325	83,325
RESIDUAL HEAT REMOVAL	0	0	0	0	397,395	397,395
POOL LINER & RACKS	0	0	0	0	2,442,058	2,442,058
CONTAMINATED CONCRETE	0	0	0	0	2,781,766	2,781,766
OTHER REACTOR BUILDING	0	0	0	0	9,095,222	9,095,222
TURBINE	0	0	0	0	19,116,146	19,116,146
NUCLEAR STEAM CONDENSATE	0	0	0	0	2,326,685	2,326,685
LOW PRESSURE FEEDWATER HEATERS	0	0	0	0	4,723,875	4,723,875
MAIN STEAM	0	0	0	0	455,082	455,082
MOISTURE SEPARATOR REHEATERS	0	0	0	0	4,582,864	4,582,864
REACTOR FEEDWATER PUMPS	0	0	0	0	2,637,647	2,637,647
HIGH PRESSURE FEEDWATER HEATERS	0	0	0	0	775,562	775,562
OTHER TG BLDG	0	0	0	0	31,131,427	31,131,427
RAD WASTE BLDG	0	0	0	0	15,415,088	15,415,088
REACTOR BLDG	0	0	0	0	1,922,880	1,922,880
TG BLDG	0	0	0	0	1,301,149	1,301,149
RAD WASTE & CONTROL	0	0	0	0	1,121,680	1,121,680
CONCENTRATOR BOTTOMS	2,088,364	3,093,750	1,701,000	0	0	6,883,114
OTHER	567,774	838,750	461,160	0	0	1,867,684
POST-TMI-2 ADDITIONS	117,533	0	0	0	0	117,533
HEAVY OBJECT SURCHARGE						0
SITE AVAILABILITY CHARGES						399,078
SUBTOTAL BWR COSTS	**3,223,975**	**8,098,750**	**4,936,680**	**777,649**	**126,787,907**	**144,224,039**
TAXES & FEES (% OF CHARGES)						749,754
TAXES & FEES ($/UNIT VOL.)						601,878
ANNUAL PERMIT FEES (3 YRS)						127,200
TOTAL BWR COSTS						**145,702,871**

(a) GTCC Material: Assumes a low-density, distributed packaging scheme and final disposal as LLW. High-density packaging, ISFSI storage, and geologic repository disposal could reduce disposal costs.

NUREG-1307

Table B-33. PWR LLW Disposition Costs Using a Combination of Non-Compact Disposal Facility and the South Carolina Disposal Facility (2012 dollars)

REFERENCE PWR COMPONENT	BASE DISPOSAL CHARGE	CASK HANDLING	CURIE SURCHARGE	LINER DOSE RATE	DOSE RATE SURCHARGE	NON-COMPACT CHARGE	DISPOSAL COST
VESSEL WALL	4,470,517	3,246,036	8,553,154	0	2,145,848	0	18,415,555
VESSEL HEAD & BOTTOM	2,848,544	3,416,880	11,260	0	0	0	6,276,684
UPPER CORE SUPPORT ASSM	268,922	341,688	5,630	0	86,055	0	702,295
UPPER SUPPORT COLUMN	248,626	341,688	56,300	0	79,560	0	726,174
UPPER CORE BARREL	118,395	170,844	450,166	0	56,830	0	796,235
UPPER CORE GRID PLATE	295,988	427,110	1,125,415	0	142,074	0	1,990,587
GUIDE TUBES	438,062	512,532	56,300	0	118,277	0	1,125,170
LOWER CORE BARREL[a]	1,894,320	2,733,504	7,202,656	0	909,274	0	12,739,754
THERMAL SHIELDS[a]	355,185	512,532	1,350,498	0	170,489	0	2,388,704
CORE SHROUD[a]	275,011	341,688	13,730,063	0	132,005	0	14,478,767
LOWER GRID PLATE[a]	295,988	427,110	2,250,830	0	142,074	0	3,116,002
LOWER SUPPORT COLUMN	75,095	85,422	225,083	0	36,046	0	421,646
LOWER CORE FORGING	815,899	939,642	1,407,500	0	391,632	0	3,554,673
MISC INTERNALS	661,440	683,376	1,126,000	0	317,491	0	2,788,307
BIO SHIELD CONCRETE	0	0	0	0	0	4,530,240	4,530,240
REACTOR CAVITY LINER	324,736	0	5,630	0	0	0	330,366
REACTOR COOLANT PUMPS	0	0	0	0	0	1,617,000	1,617,000
PRESSURIZER	0	0	0	0	0	1,386,000	1,386,000
R.Hx,EHx,SUMP PUMP, CAVITY PUMP	0	0	0	0	0	72,600	72,600
PRESSURIZER RELIEF TANK	0	0	0	0	0	217,800	217,800
SAFETY INJECTION ACCUM TANKS	0	0	0	0	0	726,000	726,000
STEAM GENERATORS	0	0	0	0	0	8,224,370	8,224,370
REACTOR COOLANT PIPING	0	0	0	0	0	598,950	598,950
REMAINING CONTAM. MATLS	0	0	0	0	0	9,548,352	9,548,352
CONTAMINATED MATRL OTHR BLD	0	0	0	0	0	86,595,647	86,595,647
FILTER CARTRIDGES	0	0	0	0	0	159,390	159,390
SPENT RESINS	1,488,240	1,708,440	4,501,660	0	714,355	0	8,412,695
COMBUSTIBLE WASTES	0	0	0	0	0	6,404,063	6,404,063
EVAPORATOR BOTTOMS	6,994,728	8,029,668	21,157,802	0	955,450	0	37,137,648
POST-TMI-2 ADDITIONS	14,036,581	0	0	0	0	0	14,036,581
SUBTOTAL PWR COSTS	**35,906,275**	**23,918,160**	**63,215,947**	**0**	**6,397,459**	**120,080,411**	**249,518,252**
ATLANTIC COMPACT COMMISSION ADMINISTRATIVE SURCHARGE							264,450
TOTAL PWR COSTS (INSIDE COMPACT)							**249,782,702**

(a) GTCC Material: Assumes a low-density, distributed packaging scheme and final disposal as LLW. High-density packaging, ISFSI storage, and geologic repository disposal could reduce disposal costs.

Table B-34. BWR LLW Disposition Costs Using a Combination of Non-Compact Disposal Facility and the South Carolina Disposal Facility (2012 dollars)

REFERENCE BWR COMPONENT	BASE DISPOSAL CHARGE	CASK HANDLING	CURIE SURCHARGE	LINER DOSE RATE	DOSE RATE SURCHARGE	NON-COMPACT CHARGE	DISPOSAL COST
STEAM SEPARATOR	275,517	2,391,816	3,151,162	0	132,248	0	5,950,743
FUEL SUPPORT & PIECES	121,259	1,195,908	394,100	0	58,204	0	1,769,471
CONTROL RODS/INCORES	361,093	683,376	1,800,664	0	173,325	0	3,018,458
CONTROL RODS GUIDES	101,506	1,025,064	56,300	0	37,557	0	1,220,428
JET PUMPS	292,675	3,416,880	4,501,660	0	140,484	0	8,351,699
TOP FUEL GUIDES	501,729	6,150,384	16,205,976	0	240,830	0	23,098,918
CORE SUPPORT PLATE	337,215	2,648,082	365,950	0	124,769	0	3,476,016
CORE SHROUD[a]	982,552	11,959,080	31,511,620	0	471,625	0	44,924,877
REACTOR VESSEL WALL	213,878	1,879,284	1,216,080	0	79,135	0	3,388,377
SAC SHIELD Neutron-Activated Matl	0	0	0	0	0	576,864	576,864
REACT. WATER REC	0	0	0	0	0	564,045	564,045
SAC SHIELD Contaminated Matl	0	0	0	0	0	1,986,976	1,986,976
OTHER PRIMARY CONTAINMENT	0	0	0	0	0	22,664,345	22,664,345
CONTAINM. ATMOSPHERIC	0	0	0	0	0	307,661	307,661
HIGH PRESSURE CORE SPRAY	0	0	0	0	0	108,963	108,963
LOW PRESSURE CORE SPRAY	0	0	0	0	0	64,096	64,096
REACTOR BLDG CLOSED COOLING	0	0	0	0	0	205,107	205,107
REACTOR CORE ISO COOLING	0	0	0	0	0	83,325	83,325
RESIDUAL HEAT REMOVAL	0	0	0	0	0	397,395	397,395
POOL LINER & RACKS	0	0	0	0	0	2,442,058	2,442,058
CONTAMINATED CONCRETE	0	0	0	0	0	2,781,766	2,781,766
OTHER REACTOR BUILDING	0	0	0	0	0	9,095,222	9,095,222
TURBINE	0	0	0	0	0	19,116,146	19,116,146
NUCLEAR STEAM CONDENSATE	0	0	0	0	0	2,326,685	2,326,685
LOW PRESSURE FEEDWATER HEATERS	0	0	0	0	0	4,723,875	4,723,875
MAIN STEAM	0	0	0	0	0	455,082	455,082
MOISTURE SEPARATOR REHEATERS	0	0	0	0	0	4,582,864	4,582,864
REACTOR FEEDWATER PUMPS	0	0	0	0	0	2,637,647	2,637,647
HIGH PRESSURE FEEDWATER HEATERS	0	0	0	0	0	775,562	775,562
OTHER TG BLDG	0	0	0	0	0	31,131,427	31,131,427
RAD WASTE BLDG	0	0	0	0	0	15,415,088	15,415,088
REACTOR BLDG	0	0	0	0	0	1,922,880	1,922,880
TG BLDG	0	0	0	0	0	1,301,149	1,301,149
RAD WASTE & CONTROL	0	0	0	0	0	1,121,680	1,121,680
CONCENTRATOR BOTTOMS	28,744,710	19,219,950	50,643,675	0	3,893,950	0	102,502,285
OTHER	7,793,010	5,210,742	539,917	0	194,186	0	13,737,856
POST-TMI-2 ADDITIONS	1,147,242	0	0	0	0	0	1,147,242
SUBTOTAL BWR COSTS	**40,872,386**	**55,780,566**	**110,387,104**	**0**	**5,546,314**	**126,787,907**	**339,374,278**
ATLANTIC COMPACT COMMISSION ADMINISTRATIVE SURCHARGE							209,349
TOTAL BWR COSTS (INSIDE COMPACT)							**339,583,627**

(a) GTCC Material: Assumes a low-density, distributed packaging scheme and final disposal as LLW. High-density packaging, ISFSI storage, and geologic repository disposal could reduce disposal costs.

Table B-35. PWR Disposition Costs Using Waste Vendors with Burial Costs at the Washington Site (2010 dollars)

REFERENCE PWR COMPONENT	VOLUME CHARGE	SHIPMENT CHARGE	CONTAINER CHARGE	CONTAINER DOSE RATE CHARGE	WASTE VENDOR CHARGE	DISPOSAL COST
VESSEL WALL	386,384	508,060	340,480	277,400	0	1,512,324
VESSEL HEAD & BOTTOM	406,720	534,800	358,400	680	0	1,300,600
UPPER CORE SUPPORT ASSM	40,672	53,480	35,840	19,400	0	149,392
UPPER SUPPORT COLUMN	40,672	53,480	35,840	19,400	0	149,392
UPPER CORE BARREL	20,336	26,740	17,920	14,600	0	79,596
UPPER CORE GRID PLATE	50,840	66,850	44,800	36,500	0	198,990
GUIDE TUBES	61,008	80,220	53,760	29,100	0	224,088
LOWER CORE BARREL[a]	325,376	427,840	286,720	233,600	0	1,273,536
THERMAL SHIELDS[a]	61,008	80,220	53,760	43,800	0	238,788
CORE SHROUD[a]	40,672	53,480	35,840	29,200	0	159,192
LOWER GRID PLATE[a]	50,840	66,850	44,800	36,500	0	198,990
LOWER SUPPORT COLUMN	10,168	13,370	8,960	7,300	0	39,798
LOWER CORE FORGING	111,848	147,070	98,560	7,248	0	364,726
MISC. INTERNALS	81,344	106,960	71,680	0	0	259,984
BIO SHIELD CONCRETE	0	0	0	0	3,981,120	3,981,120
REACTOR CAVITY LINER	52,060	13,370	35,840	0	0	101,270
REACTOR COOLANT PUMPS	0	0	0	0	1,386,000	1,386,000
PRESSURIZER	0	0	0	0	1,188,000	1,188,000
R.Hx, EHx, SUMP PUMP, CAVITY PUMP	0	0	0	0	63,800	63,800
PRESSURIZER RELIEF TANK	0	0	0	0	191,400	191,400
SAFETY INJECTION ACCUM. TANKS	0	0	0	0	638,000	638,000
STEAM GENERATORS	0	0	0	0	7,049,460	7,049,460
REACTOR COOLANT PIPING	0	0	0	0	526,350	526,350
REMAINING CONTAM. MATLS	0	0	0	0	8,390,976	8,390,976
CONTAM. MATL OTHER BLDG	0	0	0	0	76,099,205	76,099,205
FILTER CARTRIDGES	0	0	0	0	138,600	138,600
SPENT RESINS	203,360	267,400	179,200	0	0	649,960
COMBUSTIBLE WASTES	0	0	0	0	5,568,750	5,568,750
EVAPORATOR BOTTOMS	955,792	1,256,780	842,240	0	0	3,054,812
POST-TMI-2 ADDITIONS	1,582,446	0	0	0	0	1,582,446
HEAVY OBJECT SURCHARGE						0
SITE AVAILABILITY CHARGES						387,039
SUBTOTAL PWR COSTS	**4,481,546**	**3,756,970**	**2,544,640**	**754,728**	**105,221,661**	**117,146,584**
TAXES & FEES (% OF CHARGES)						512,772
TAXES & FEES ($/UNIT VOL.)						725,794
ANNUAL PERMIT FEES (3 YRS)						127,200
TOTAL PWR COSTS						**118,512,349**

[a] GTCC Material: Assumes a low-density, distributed packaging scheme and final disposal as LLW. High-density packaging, ISFSI storage, and geologic repository disposal could reduce disposal costs.

NUREG-1307

Table B-36. BWR Disposition Costs Using Waste Vendors with Burial Costs at the Washington Site (2010 dollars)

REFERENCE BWR COMPONENT	VOLUME CHARGE	SHIPMENT CHARGE	CONTAINER CHARGE	CONTAINER DOSE RATE CHARGE	VENDOR CHARGE	DISPOSAL COST
STEAM SEPARATOR	35,908	187,180	250,880	754,728	0	1,228,696
FUEL SUPPORT & PIECES	17,954	93,590	125,440	0	0	236,984
CONTROL RODS/INCORES	53,862	106,960	71,680	0	0	232,502
CONTROL RODS GUIDES	14,363	80,220	107,520	0	0	202,103
JET PUMPS	50,271	267,400	358,400	0	0	676,071
TOP FUEL GUIDES	86,179	962,640	645,120	0	0	1,693,939
CORE SUPPORT PLATE	39,499	213,920	277,760	0	0	531,179
CORE SHROUD[a]	168,767	1,871,800	1,254,400	0	0	3,294,967
REACTOR VESSEL WALL	28,726	267,400	197,120	0	0	493,246
SAC SHIELD NEUTRON ACTIV. MATL.	0	0	0	0	506,941	506,941
REACTOR WATER REC	0	0	0	0	495,676	495,676
SAC SHIELD CONTAMINATED MTL.	0	0	0	0	1,746,130	1,746,130
OTHER PRIMARY CONTAINMENT	0	0	0	0	19,917,152	19,917,152
CONTAINMENT ATMOSPHERIC	0	0	0	0	270,369	270,369
HIGH PRESSURE CORE SPRAY	0	0	0	0	95,756	95,756
LOW PRESSURE CORE SPRAY	0	0	0	0	56,327	56,327
REACTOR BLDG CLOSED COOLING	0	0	0	0	180,246	180,246
REACTOR CORE ISO COOLING	0	0	0	0	73,225	73,225
RESIDUAL HEAT REMOVAL	0	0	0	0	349,226	349,226
POOL LINER & RACKS	0	0	0	0	2,146,051	2,146,051
CONTAMINATED CONCRETE	0	0	0	0	2,444,583	2,444,583
OTHER REACTOR BUILDING	0	0	0	0	7,992,771	7,992,771
TURBINE	0	0	0	0	16,385,268	16,385,268
NUCLEAR STEAM CONDENSATE	0	0	0	0	2,044,662	2,044,662
LOW PRESSURE FEEDWATER HEATERS	0	0	0	0	4,151,284	4,151,284
MAIN STEAM	0	0	0	0	399,920	399,920
MOISTURE SEPARATOR REHEATERS	0	0	0	0	4,027,365	4,027,365
REACTOR FEEDWATER PUMPS	0	0	0	0	2,260,841	2,260,841
HIGH PRESSURE FEEDWATER HEATERS	0	0	0	0	681,554	681,554
OTHER TG BLDG	0	0	0	0	27,357,920	27,357,920
RAD WASTE BLDG	0	0	0	0	13,546,592	13,546,592
REACTOR BLDG	0	0	0	0	1,689,804	1,689,804
TG BLDG	0	0	0	0	1,143,434	1,143,434
RAD WASTE & CONTROL	0	0	0	0	985,719	985,719
CONCENTRATOR BOTTOMS	2,298,105	3,008,250	2,016,000	0	0	7,322,355
OTHER	624,797	815,570	546,560	0	0	1,986,927
POST-TMI-2 ADDITIONS	129,337	0	0	0	0	129,337
HEAVY OBJECT SURCHARGE						0
SITE AVAILABILITY CHARGES						387,039
SUBTOTAL BWR COSTS	**3,547,768**	**7,874,930**	**5,850,880**	**754,728**	**110,948,815**	**129,364,160**
TAXES & FEES (% OF CHARGES)						791,860
TAXES & FEES ($/UNIT VOL.)						601,878
ANNUAL PERMIT FEES (3 YRS)						127,200
TOTAL BWR COSTS						**130,885,098**

(a) GTCC Material: Assumes a low-density, distr buted packaging scheme and final disposal as LLW. High-density packaging, ISFSI storage, and geologic repository disposal could reduce disposal costs.

Table B-37. PWR Disposition Costs Using Waste Vendors with Burial Costs at the South Carolina Site (2010 dollars)

REFERENCE PWR COMPONENT	BASE DISPOSAL CHARGE	CASK HANDLING	CURIE SURCHARGE	LINER DOSE RATE	DOSE RATE SURCHARGE	VENDOR CHARGES	DISPOSAL COST
VESSEL WALL	3,986,350	2,894,422	7,626,676	0	1,913,448	0	16,420,895
VESSEL HEAD & BOTTOM	2,540,054	3,046,760	10,040	0	0	0	5,596,854
UPPER CORE SUPPORT ASSM	239,799	304,676	5,020	0	76,736	0	626,230
UPPER SUPPORT COLUMN	221,701	304,676	50,200	0	70,944	0	647,521
UPPER CORE BARREL	105,570	152,338	401,404	0	50,674	0	709,986
UPPER CORE GRID PLATE	263,925	380,845	1,003,510	0	126,684	0	1,774,964
GUIDE TUBES	390,609	457,014	50,200	0	105,464	0	1,003,287
LOWER CORE BARREL[a]	1,689,120	2,437,408	6,422,464	0	810,778	0	11,359,770
THERMAL SHIELDS[a]	316,710	457,014	1,204,212	0	152,021	0	2,129,957
CORE SHROUD[a]	245,228	304,676	12,242,822	0	117,709	0	12,910,435
LOWER GRID PLATE[a]	263,925	380,845	2,007,020	0	126,684	0	2,778,474
LOWER SUPPORT COLUMN	66,963	76,169	200,702	0	32,142	0	375,976
LOWER CORE FORGING	727,540	837,859	1,255,000	0	349,219	0	3,169,618
MISC. INTERNALS	589,840	609,352	1,004,000	0	283,123	0	2,486,315
BIO SHIELD CONCRETE	0	0	0	0	0	3,981,120	3,981,120
REACTOR CAVITY LINER	289,568	0	5,020	0	0	0	294,588
REACTOR COOLANT PUMPS	0	0	0	0	0	1,386,000	1,386,000
PRESSURIZER	0	0	0	0	0	1,188,000	1,188,000
R.Hx, EHx, SUMP PUMP, CAVITY PUMP	0	0	0	0	0	63,800	63,800
PRESSURIZER RELIEF TANK	0	0	0	0	0	191,400	191,400
SAFETY INJECTION ACCUM. TANKS	0	0	0	0	0	638,000	638,000
STEAM GENERATORS	0	0	0	0	0	7,049,460	7,049,460
REACTOR COOLANT PIPING	0	0	0	0	0	526,350	526,350
REMAINING CONTAM. MATLS	0	0	0	0	0	8,390,976	8,390,976
CONTAM. MATL OTHER BLDG	0	0	0	0	0	76,099,205	76,099,205
FILTER CARTRIDGES	0	0	0	0	0	138,600	138,600
SPENT RESINS	1,327,140	1,523,380	4,014,040	0	637,027	0	7,501,587
COMBUSTIBLE WASTES	0	0	0	0	0	5,568,750	5,568,750
EVAPORATOR BOTTOMS	6,237,558	7,159,886	18,865,988	0	852,024	0	33,115,456
POST-TMI-2 ADDITIONS	12,516,387	0	0	0	0	0	12,516,387
SUBTOTAL PWR COSTS	**32,017,985**	**21,327,320**	**56,368,318**	**0**	**5,704,677**	**105,221,661**	**220,639,960**
ATLANTIC COMPACT COMMISSION ADMINISTRATIVE SURCHARGE							264,450
TOTAL PWR COSTS (INSIDE COMPACT)							**220,904,410**

(a) GTCC Material: Assumes a low-density, distributed packaging scheme and final disposal as LLW. High-density packaging, ISFSI storage, and geologic repository disposal could reduce disposal costs.

Table B-38. BWR Disposition Costs Using Waste Vendors with Burial Costs at the South Carolina Site (2010 dollars)

REFERENCE BWR COMPONENT	BASE DISPOSAL CHARGE	CASK HANDLING	CURIE SURCHARGE	LINER DOSE RATE	DOSE RATE SURCHARGE	VENDOR CHARGES	DISPOSAL COST
STEAM SEPARATOR	245,672	2,132,732	2,809,828	0	117,923	0	5,306,155
FUEL SUPPORT & PIECES	108,127	1,066,366	351,400	0	51,901	0	1,577,794
CONTROL RODS/INCORES	322,005	609,352	1,605,616	0	154,563	0	2,691,536
CONTROL RODS GUIDES	90,511	914,028	50,200	0	33,489	0	1,088,228
JET PUMPS	260,971	3,046,760	4,014,040	0	125,266	0	7,447,038
TOP FUEL GUIDES	447,379	5,484,168	14,450,544	0	214,742	0	20,596,834
CORE SUPPORT PLATE	300,712	2,361,239	326,300	0	111,263	0	3,099,514
CORE SHROUD[a]	876,118	10,663,660	28,098,280	0	420,537	0	40,058,595
REACTOR VESSEL WALL	190,716	1,675,718	1,084,320	0	70,565	0	3,021,319
SAC SHIELD NEUTRON-ACTIV. MATL	0	0	0	0	0	506,941	506,941
REACT. WATER REC	0	0	0	0	0	495,676	495,676
SAC SHIELD-CONTAMINATED MATL	0	0	0	0	0	1,746,130	1,746,130
OTHER PRIMARY CONTAINMENT	0	0	0	0	0	19,917,152	19,917,152
CONTAINM. ATMOSPHERIC	0	0	0	0	0	270,369	270,369
HIGH PRESSURE CORE SPRAY	0	0	0	0	0	95,756	95,756
LOW PRESSURE CORE SPRAY	0	0	0	0	0	56,327	56,327
REACTOR BLDG CLOSED COOLING	0	0	0	0	0	180,246	180,246
REACTOR CORE ISO COOLING	0	0	0	0	0	73,225	73,225
RESIDUAL HEAT REMOVAL	0	0	0	0	0	349,226	349,226
POOL LINER & RACKS	0	0	0	0	0	2,146,051	2,146,051
CONTAMINATED CONCRETE	0	0	0	0	0	2,444,583	2,444,583
OTHER REACTOR BUILDING	0	0	0	0	0	7,992,771	7,992,771
TURBINE	0	0	0	0	0	16,385,268	16,385,268
NUCLEAR STEAM CONDENSATE	0	0	0	0	0	2,044,662	2,044,662
LOW PRESSURE FEEDWATER HEATERS	0	0	0	0	0	4,151,284	4,151,284
MAIN STEAM	0	0	0	0	0	399,920	399,920
MOISTURE SEPARATOR REHEATERS	0	0	0	0	0	4,027,365	4,027,365
REACTOR FEEDWATER PUMPS	0	0	0	0	0	2,260,841	2,260,841
HIGH PRESSURE FEEDWATER HEATERS	0	0	0	0	0	681,554	681,554
OTHER TG BLDG	0	0	0	0	0	27,357,920	27,357,920
RAD WASTE BLDG	0	0	0	0	0	13,546,592	13,546,592
REACTOR BLDG	0	0	0	0	0	1,689,804	1,689,804
TG BLDG	0	0	0	0	0	1,143,434	1,143,434
RAD WASTE & CONTROL	0	0	0	0	0	985,719	985,719
CONCENTRATOR BOTTOMS	25,631,595	17,138,025	45,157,950	0	3,472,227	0	91,399,797
OTHER	6,949,010	4,646,309	481,418	0	173,156	0	12,249,893
POST-TMI-2 ADDITIONS	1,022,993	0	0	0	0	0	1,022,993
SUBTOTAL BWR COSTS	36,445,811	49,738,357	98,429,896	0	4,945,631	110,948,815	300,508,509
ATLANTIC COMPACT COMMISSION ADMINISTRATIVE SURCHARGE							209,349
TOTAL BWR COSTS (INSIDE COMPACT)							300,717,858

(a) GTCC Material: Assumes a low-density, distributed packaging scheme and final disposal as LLW. High-density packaging, ISFSI storage, and geologic repository disposal could reduce disposal costs.

Table B-39. PWR Disposition Costs Using Waste Vendors with Burial Costs at the Washington Site (2008 dollars)

REFERENCE PWR COMPONENT	VOLUME CHARGE	SHIPMENT CHARGE	CONTAINER CHARGE	CONTAINER DOSE RATE CHARGE	WASTE VENDOR CHARGE	DISPOSAL COST
VESSEL WALL	375,060	560,120	269,040	2,869,000	0	4,073,220
VESSEL HEAD & BOTTOM	394,800	589,600	283,200	7,080	0	1,274,680
UPPER CORE SUPPORT ASSM	39,480	58,960	28,320	201,600	0	328,360
UPPER SUPPORT COLUMN	39,480	58,960	28,320	201,600	0	328,360
UPPER CORE BARREL	19,740	29,480	14,160	151,000	0	214,380
UPPER CORE GRID PLATE	49,350	73,700	35,400	377,500	0	535,950
GUIDE TUBES	59,220	88,440	42,480	302,400	0	492,540
LOWER CORE BARREL[(a)]	315,840	471,680	226,560	2,416,000	0	3,430,080
THERMAL SHIELDS[(a)]	59,220	88,440	42,480	453,000	0	643,140
CORE SHROUD[(a)]	39,480	58,960	28,320	302,000	0	428,760
LOWER GRID PLATE[(a)]	49,350	73,700	35,400	377,500	0	535,950
LOWER SUPPORT COLUMN	9,870	14,740	7,080	75,500	0	107,190
LOWER CORE FORGING	108,570	162,140	77,880	830,500	0	1,179,090
MISC. INTERNALS	78,960	117,920	56,640	604,000	0	857,520
BIO SHIELD CONCRETE	0	0	0	0	3,193,693	3,193,693
REACTOR CAVITY LINER	50,534	14,740	28,320	708	0	94,302
REACTOR COOLANT PUMPS	0	0	0	0	1,231,619	1,231,619
PRESSURIZER	0	0	0	0	319,369	319,369
R.Hx, EHx, SUMP PUMP, CAVITY PUMP	0	0	0	0	19,326	19,326
PRESSURIZER RELIEF TANK	0	0	0	0	44,548	44,548
SAFETY INJECTION ACCUM. TANKS	0	0	0	0	501,164	501,164
STEAM GENERATORS	0	0	0	0	4,507,202	4,507,202
REACTOR COOLANT PIPING	0	0	0	0	363,586	363,586
REMAINING CONTAM. MATLS	0	0	0	0	6,427,512	6,427,512
CONTAMINATED MATRL OTHR BLD	0	0	0	0	49,447,064	49,447,064
FILTER CARTRIDGES	0	0	0	0	88,441	88,441
SPENT RESINS	197,400	294,800	141,600	1,510,000	0	2,143,800
COMBUSTIBLE WASTES	0	0	0	0	884,407	884,407
EVAPORATOR BOTTOMS	927,780	1,385,560	665,520	2,231,879	0	5,210,739
POST-TMI-2 ADDITIONS	1,536,068	0	0	0	0	1,536,068
HEAVY OBJECT SURCHARGE						0
SITE AVAILABILITY CHARGES, (3 YRS)						374,400
SUBTOTAL PWR COSTS	**4,350,203**	**4,141,940**	**2,010,720**	**12,911,267**	**67,027,931**	**90,816,460**
TAXES & FEES (% OF CHARGES)						1,022,907
TAXES & FEES ($/UNIT VOL)						725,794
ANNUAL PERMIT FEES (3 YRS)						127,200
TOTAL PWR COSTS						**92,692,361**

(a) GTCC Material: Assumes a low-density, distributed packaging scheme and final disposal as LLW. High-density packaging, ISFSI storage, and geologic repository disposal could reduce disposal costs.

Table B-40. BWR Disposition Costs Using Waste Vendors with Burial Costs at the Washington Site (2008 dollars)

REFERENCE BWR COMPONENT	VOLUME CHARGE	SHIPMENT CHARGE	CONTAINER CHARGE	CONTAINER DOSE RATE CHARGE	WASTE VENDOR CHARGE	DISPOSAL COST
STEAM SEPARATOR	34,841	206,360	198,240	35,504,000	0	35,943,441
FUEL SUPPORT & PIECES	17,470	103,180	99,120	1,057,000	0	1,276,770
CONTROL RODS/INCORES	52,311	117,920	56,640	10,144,000	0	10,370,871
CONTROL RODS GUIDES	13,917	88,440	84,960	906,000	0	1,093,317
JET PUMPS	48,857	294,800	283,200	50,720,000	0	51,346,857
TOP FUEL GUIDES	83,698	1,061,280	509,760	91,296,000	0	92,950,738
CORE SUPPORT PLATE	38,394	235,840	219,480	2,340,500	0	2,834,214
CORE SHROUD[a]	163,842	2,063,600	991,200	177,520,000	0	180,738,642
REACTOR VESSEL WALL	27,932	294,800	155,760	1,661,000	0	2,139,492
SAC SHIELD	0	0	0	0	1,632,589	1,632,589
REACTOR WATER REC	0	0	0	0	523,267	523,267
SAC SHIELD	0	0	0	0	4,227,912	4,227,912
OTHER PRIMARY CONTAINMENT	0	0	0	0	16,944,058	16,944,058
CONTAINMENT ATMOSPHERIC	0	0	0	0	68,206	68,206
HIGH PRESSURE CORE SPRAY	0	0	0	0	166,423	166,423
LOW PRESSURE CORE SPRAY	0	0	0	0	60,021	60,021
REACTOR BLDG CLOSED COOLING	0	0	0	0	163,258	163,258
REACTOR CORE ISO COOLING	0	0	0	0	52,999	52,999
RESIDUAL HEAT REMOVAL	0	0	0	0	502,003	502,003
POOL LINER & RACKS	0	0	0	0	2,080,944	2,080,944
CONTAMINATED CONCRETE	0	0	0	0	2,975,264	2,975,264
OTHER REACTOR BUILDING	0	0	0	0	3,708,077	3,708,077
TURBINE	0	0	0	0	9,980,248	9,980,248
NUCLEAR STEAM CONDENSATE	0	0	0	0	1,319,092	1,319,092
LOW PRESSURE FEEDWATER HEATERS	0	0	0	0	4,429,584	4,429,584
MAIN STEAM	0	0	0	0	193,708	193,708
MOISTURE SEPARATOR REHEATERS	0	0	0	0	2,504,756	2,504,756
REACTOR FEEDWATER PUMPS	0	0	0	0	658,856	658,856
HIGH PRESSURE FEEDWATER HEATERS	0	0	0	0	882,075	882,075
OTHER TG BLDG	0	0	0	0	19,145,326	19,145,326
RAD WASTE BLDG	0	0	0	0	6,539,338	6,539,338
REACTOR BLDG	0	0	0	0	4,664,920	4,664,920
TG BLDG	0	0	0	0	3,069,026	3,069,026
RAD WASTE & CONTROL	0	0	0	0	2,823,504	2,823,504
CONCENTRATOR BOTTOMS	2,220,750	3,316,500	1,593,000	5,296,145	0	12,426,395
OTHER	602,070	899,140	431,880	246,454	0	2,179,544
POST-TMI-2 ADDITIONS	125,546	0	0	0	0	125,546
HEAVY OBJECT SURCHARGE						0
SITE AVAILABILITY CHARGES, (3.5 YRS)						499,200
SUBTOTAL BWR COSTS	**3,429,628**	**8,681,860**	**4,623,240**	**376,691,099**	**89,315,455**	**483,240,482**
TAXES & FEES (% OF CHARGES)						16,938,776
TAXES & FEES ($/UNIT VOL.)						599,403
ANNUAL PERMIT FEES (3.5 YRS)						169,600
TOTAL BWR COSTS						**500,948,261**

(a) GTCC Material: Assumes a low-density, distributed packaging scheme and final disposal as LLW. High-density packaging, ISFSI storage, and geologic repository disposal could reduce disposal costs.

NUREG-1307

Table B-41. PWR Disposition Costs Using Waste Vendors with Burial Costs at the South Carolina Site Atlantic Compact (2008 dollars)

REFERENCE PWR COMPONENT	BASE DISPOSAL CHARGE	CASK HANDLING	CURIE SURCHARGE	DOSE RATE SURCHARGE	WASTE VENDOR CHARGE	DISPOSAL COST
VESSEL WALL	3,682,407	2,673,832	7,052,800	1,767,555	0	15,176,594
VESSEL HEAD & BOTTOM	2,346,371	2,814,560	9,280	0	0	5,170,211
UPPER CORE SUPPORT ASSM	221,514	281,456	4,640	70,884	0	578,494
UPPER SUPPORT COLUMN	204,796	281,456	46,400	65,535	0	598,186
UPPER CORE BARREL	97,524	140,728	371,200	46,812	0	656,264
UPPER CORE GRID PLATE	243,810	351,820	928,000	117,029	0	1,640,659
GUIDE TUBES	360,839	422,184	46,400	97,426	0	926,849
LOWER CORE BARREL[a]	1,560,384	2,251,648	5,939,200	748,984	0	10,500,216
THERMAL SHIELDS[a]	292,572	422,184	1,113,600	140,435	0	1,968,791
CORE SHROUD[a]	226,529	281,456	11,321,600	108,734	0	11,938,319
LOWER GRID PLATE[a]	243,810	351,820	1,856,000	117,029	0	2,568,659
LOWER SUPPORT COLUMN	61,857	70,364	185,600	29,691	0	347,512
LOWER CORE FORGING	672,064	774,004	1,160,000	322,591	0	2,928,658
MISC. INTERNALS	544,880	562,912	928,000	261,542	0	2,297,334
BIO SHIELD CONCRETE	0	0	0	0	3,193,693	3,193,693
REACTOR CAVITY LINER	267,488	0	4,640	0	0	272,128
REACTOR COOLANT PUMPS	0	0	0	0	1,231,619	1,231,619
PRESSURIZER	0	0	0	0	319,369	319,369
R.Hx, EHx, SUMP PUMP, CAVITY PUMP	0	0	0	0	19,326	19,326
PRESSURIZER RELIEF TANK	0	0	0	0	44,548	44,548
SAFETY INJECTION ACCUM. TANKS	0	0	0	0	501,164	501,164
STEAM GENERATORS	0	0	0	0	4,507,202	4,507,202
REACTOR COOLANT PIPING	0	0	0	0	363,586	363,586
REMAINING CONTAM. MATLS	0	0	0	0	6,427,512	6,427,512
CONTAM. MATL OTHER BLDG	0	0	0	0	49,447,064	49,447,064
FILTER CARTRIDGES	0	0	0	0	88,441	88,441
SPENT RESINS	1,225,980	1,407,280	3,712,000	588,470	0	6,933,730
COMBUSTIBLE WASTES	0	0	0	0	884,407	884,407
EVAPORATOR BOTTOMS	5,762,106	6,614,216	17,446,400	787,079	0	30,609,801
POST-TMI-2 ADDITIONS	11,562,064	0	0	0	0	11,562,064
SITE ACCESS FEES (3 YRS)						0
SUBTOTAL PWR COSTS	**29,576,993**	**19,701,920**	**52,125,760**	**5,269,796**	**67,027,931**	**173,702,400**
BARNWELL COUNTY BUSINESS TAX						0
ATLANTIC COMPACT SURCHARGE (INSIDE COMPACT)						3,883,482
TOTAL PWR COSTS (INSIDE COMPACT)						**177,585,882**

(a) GTCC Material: Assumes a low-density, distributed packaging scheme and final disposal as LLW. High-density packaging, ISFSI storage, and geologic repository disposal could reduce disposal costs.

Table B-42. BWR Disposition Costs Using Waste Vendors with Burial Costs at the South Carolina Site Atlantic Compact (2008 dollars)

REFERENCE BWR COMPONENT	BASE DISPOSAL CHARGE	CASK HANDLING	CURIE SURCHARGE	DOSE RATE SURCHARGE	WASTE VENDOR CHARGE	DISPOSAL COST
STEAM SEPARATOR	226,342	1,970,192	2,598,400	108,644	0	4,903,579
FUEL SUPPORT & PIECES	99,706	985,096	324,800	47,859	0	1,457,461
CONTROL RODS/INCORES	296,851	562,912	1,484,800	142,488	0	2,487,051
CONTROL RODS GUIDES	83,437	844,368	46,400	30,872	0	1,005,077
JET PUMPS	241,372	2,814,560	3,712,000	115,859	0	6,883,790
TOP FUEL GUIDES	413,502	5,066,208	13,363,200	198,481	0	19,041,391
CORE SUPPORT PLATE	277,208	2,181,284	301,600	102,567	0	2,862,659
CORE SHROUD[a]	809,449	9,850,960	25,984,000	388,536	0	37,032,945
REACTOR VESSEL WALL	176,108	1,548,008	1,002,240	65,160	0	2,791,516
SAC SHIELD	0	0	0	0	1,632,589	1,632,589
REACTOR WATER REC	0	0	0	0	523,267	523,267
SAC SHIELD	0	0	0	0	4,227,912	4,227,912
OTHER PRIMARY CONTAINMENT	0	0	0	0	16,944,058	16,944,058
CONTAINMENT ATMOSPHERIC	0	0	0	0	68,206	68,206
HIGH PRESSURE CORE SPRAY	0	0	0	0	166,423	166,423
LOW PRESSURE CORE SPRAY	0	0	0	0	60,021	60,021
REACTOR BLDG CLOSED COOLING	0	0	0	0	163,258	163,258
REACTOR CORE ISO COOLING	0	0	0	0	52,999	52,999
RESIDUAL HEAT REMOVAL	0	0	0	0	502,003	502,003
POOL LINER & RACKS	0	0	0	0	2,080,944	2,080,944
CONTAMINATED CONCRETE	0	0	0	0	2,975,264	2,975,264
OTHER REACTOR BUILDING	0	0	0	0	3,708,077	3,708,077
TURBINE	0	0	0	0	9,980,248	9,980,248
NUCLEAR STEAM CONDENSATE	0	0	0	0	1,319,092	1,319,092
LOW PRESSURE FEEDWATER HEATERS	0	0	0	0	4,429,584	4,429,584
MAIN STEAM	0	0	0	0	193,708	193,708
MOISTURE SEPARATOR REHEATERS	0	0	0	0	2,504,756	2,504,756
REACTOR FEEDWATER PUMPS	0	0	0	0	658,856	658,856
HIGH PRESSURE FEEDWATER HEATERS	0	0	0	0	882,075	882,075
OTHER TG BLDG	0	0	0	0	19,145,326	19,145,326
RAD WASTE BLDG	0	0	0	0	6,539,338	6,539,338
REACTOR BLDG	0	0	0	0	4,664,920	4,664,920
TG BLDG	0	0	0	0	3,069,026	3,069,026
RAD WASTE & CONTROL	0	0	0	0	2,823,504	2,823,504
CONCENTRATOR BOTTOMS	23,677,260	15,831,900	41,760,000	3,207,479	0	84,476,639
OTHER	6,419,168	4,292,204	444,976	159,953	0	11,316,301
POST-TMI-2 ADDITIONS	944,994	0	0	0	0	944,994
SITE ACCESS FEES (3.5 YRS)						0
SUBTOTAL BWR COSTS	**33,665,397**	**45,947,692**	**91,022,416**	**4,567,898**	**89,315,455**	**264,518,858**
BARNWELL COUNTY BUSINESS TAX						0
ATLANTIC COMPACT SURCHARGE (INSIDE COMPACT)						4,021,086
TOTAL BWR COSTS (INSIDE COMPACT)						**268,539,944**

(a) GTCC Material: Assumes a low-density, distrbuted packaging scheme and final disposal as LLW. High-density packaging, ISFSI storage, and geologic repository disposal could reduce disposal costs.

Table B-43. PWR Disposition Costs Using Waste Vendors with Burial Costs at the Washington Site (2006 dollars)

REFERENCE PWR COMPONENT	VOLUME CHARGE	SHIPMENT CHARGE	CONTAINER CHARGE	CONTAINER DOSE RATE CHARGE	WASTE VENDOR CHARGE	DISPOSAL COST
VESSEL WALL	330,220	460,180	230,280	1,014,600	0	2,035,280
VESSEL HEAD & BOTTOM	347,600	484,400	242,400	2,520	0	1,076,920
UPPER CORE SUPPORT ASSM	34,760	48,440	24,240	71,200	0	178,640
UPPER SUPPORT COLUMN	34,760	48,440	24,240	71,200	0	178,640
UPPER CORE BARREL	17,380	24,220	12,120	53,400	0	107,120
UPPER CORE GRID PLATE	43,450	60,550	30,300	133,500	0	267,800
GUIDE TUBES	52,140	72,660	36,360	106,800	0	267,960
LOWER CORE BARREL[a]	278,080	387,520	193,920	854,400	0	1,713,920
THERMAL SHIELDS[a]	52,140	72,660	36,360	160,200	0	321,360
CORE SHROUD[a]	34,760	48,440	24,240	106,800	0	214,240
LOWER GRID PLATE[a]	43,450	60,550	30,300	133,500	0	267,800
LOWER SUPPORT COLUMN	8,690	12,110	6,060	26,700	0	53,560
LOWER CORE FORGING	95,590	133,210	66,660	293,700	0	589,160
MISC. INTERNALS	69,520	96,880	48,480	213,600	0	428,480
BIO SHIELD CONCRETE	0	0	0	0	2,571,846	2,571,846
REACTOR CAVITY LINER	44,493	12,110	24,240	252	0	81,095
REACTOR COOLANT PUMPS	0	0	0	0	991,810	991,810
PRESSURIZER	0	0	0	0	257,185	257,185
R.Hx, EHx, SUMP PUMP, CAVITY PUMP	0	0	0	0	15,563	15,563
PRESSURIZER RELIEF TANK	0	0	0	0	35,874	35,874
SAFETY INJECTION ACCUM. TANKS	0	0	0	0	403,582	403,582
STEAM GENERATORS	0	0	0	0	3,629,601	3,629,601
REACTOR COOLANT PIPING	0	0	0	0	292,792	292,792
REMAINING CONTAM. MATLS	0	0	0	0	5,176,006	5,176,006
CONTAM. MATL OTHER BLDG	0	0	0	0	39,819,187	39,819,187
FILTER CARTRIDGES	0	0	0	0	71,220	71,220
SPENT RESINS	173,800	242,200	121,200	534,000	0	1,071,200
COMBUSTIBLE WASTES	0	0	0	0	712,204	712,204
EVAPORATOR BOTTOMS	816,860	1,138,340	569,640	790,701	0	3,315,541
POST-TMI-2 ADDITIONS	1,352,425	0	0	0	0	1,352,425
HEAVY OBJECT SURCHARGE						0
SITE AVAILABILITY CHARGES, (3 YRS)						401,727
SUBTOTAL PWR COSTS	**3,830,118**	**3,402,910**	**1,721,040**	**4,567,073**	**53,976,869**	**67,899,737**
TAXES & FEES (% OF CHARGES)						598,683
TAXES & FEES ($/UNIT VOL)						725,794
ANNUAL PERMIT FEES (3 YRS)						127,200
TOTAL PWR COSTS						**69,351,414**

(a) GTCC Material: Assumes a low-density, distributed packaging scheme and final disposal as LLW. High-density packaging, ISFSI storage, and geologic repository disposal could reduce disposal costs.

NUREG-1307

Table B-44. BWR Disposition Costs Using Waste Vendors with Burial Costs at the Washington Site (2006 dollars)

REFERENCE BWR COMPONENT	VOLUME CHARGE	SHIPMENT CHARGE	CONTAINER CHARGE	CONTAINER DOSE RATE CHARGE	WASTE VENDOR CHARGE	DISPOSAL COST
STEAM SEPARATOR	30,676	169,540	169,680	12,555,200	0	12,925,096
FUEL SUPPORT & PIECES	15,381	84,770	84,840	373,800	0	558,791
CONTROL RODS/INCORES	46,057	96,880	48,480	3,587,200	0	3,778,617
CONTROL RODS GUIDES	12,253	72,660	72,720	320,400	0	478,033
JET PUMPS	43,016	242,200	242,400	17,936,000	0	18,463,616
TOP FUEL GUIDES	73,691	871,920	436,320	32,284,800	0	33,666,731
CORE SUPPORT PLATE	33,804	193,760	187,860	827,700	0	1,243,124
CORE SHROUD[a]	144,254	1,695,400	848,400	62,776,000	0	65,464,054
REACTOR VESSEL WALL	24,593	242,200	133,320	587,400	0	987,513
SAC SHIELD	0	0	0	0	1,115,496	1,115,496
REACTOR WATER REC	0	0	0	0	357,532	357,532
SAC SHIELD	0	0	0	0	2,888,796	2,888,796
OTHER PRIMARY CONTAINMENT	0	0	0	0	11,577,329	11,577,329
CONTAINMENT ATMOSPHERIC	0	0	0	0	46,603	46,603
HIGH PRESSURE CORE SPRAY	0	0	0	0	113,712	113,712
LOW PRESSURE CORE SPRAY	0	0	0	0	41,011	41,011
REACTOR BLDG CLOSED COOLING	0	0	0	0	111,549	111,549
REACTOR CORE ISO COOLING	0	0	0	0	36,212	36,212
RESIDUAL HEAT REMOVAL	0	0	0	0	343,003	343,003
POOL LINER & RACKS	0	0	0	0	1,421,842	1,421,842
CONTAMINATED CONCRETE	0	0	0	0	2,032,902	2,032,902
OTHER REACTOR BUILDING	0	0	0	0	2,533,609	2,533,609
TURBINE	0	0	0	0	6,819,182	6,819,182
NUCLEAR STEAM CONDENSATE	0	0	0	0	901,293	901,293
LOW PRESSURE FEEDWATER HEATERS	0	0	0	0	3,026,592	3,026,592
MAIN STEAM	0	0	0	0	132,355	132,355
MOISTURE SEPARATOR REHEATERS	0	0	0	0	1,711,419	1,711,419
REACTOR FEEDWATER PUMPS	0	0	0	0	450,175	450,175
HIGH PRESSURE FEEDWATER HEATERS	0	0	0	0	602,694	602,694
OTHER TG BLDG	0	0	0	0	13,081,385	13,081,385
RAD WASTE BLDG	0	0	0	0	4,468,119	4,468,119
REACTOR BLDG	0	0	0	0	3,187,390	3,187,390
TG BLDG	0	0	0	0	2,096,967	2,096,967
RAD WASTE & CONTROL	0	0	0	0	1,929,210	1,929,210
CONCENTRATOR BOTTOMS	1,955,250	2,724,750	1,363,500	1,876,335	0	7,919,835
OTHER	530,090	738,710	369,660	87,766	0	1,726,226
POST-TMI-2 ADDITIONS	110,537	0	0	0	0	110,537
HEAVY OBJECT SURCHARGE						0
SITE AVAILABILITY CHARGES (3.5 YRS)						535,636
SUBTOTAL BWR COSTS	**3,019,601**	**7,132,790**	**3,957,180**	**133,212,601**	**61,026,373**	**208,884,182**
TAXES & FEES (% OF CHARGES)						6,357,886
TAXES & FEES ($/UNIT VOL.)						599,403
ANNUAL PERMIT FEES (3.5 YRS)						169,600
TOTAL BWR COSTS						**216,011,070**

(a) GTCC Material: Assumes a low-density, distributed packaging scheme and final disposal as LLW. High-density packaging, ISFSI storage, and geologic repository disposal could reduce disposal costs.

Table B-45. PWR Disposition Costs Using Waste Vendors with Burial Costs at the South Carolina Site Atlantic Compact (2006 dollars)

REFERENCE PWR COMPONENT	BASE DISPOSAL CHARGE	CASK HANDLING	CURIE SURCHARGE	DOSE RATE SURCHARGE	WASTE VENDOR CHARGE	DISPOSAL COST
VESSEL WALL	3,344,560	2,428,580	6,399,200	1,605,389	0	13,777,729
VESSEL HEAD & BOTTOM	2,131,074	2,556,400	8,420	0	0	4,695,894
UPPER CORE SUPPORT ASSM	201,188	255,640	4,210	64,380	0	525,418
UPPER SUPPORT COLUMN	186,004	255,640	42,100	59,521	0	543,265
UPPER CORE BARREL	88,578	127,820	336,800	42,517	0	595,715
UPPER CORE GRID PLATE	221,445	319,550	842,000	106,294	0	1,489,289
GUIDE TUBES	327,739	383,460	42,100	88,489	0	841,788
LOWER CORE BARREL[a]	1,417,248	2,045,120	5,388,800	680,279	0	9,531,447
THERMAL SHIELDS[a]	265,734	383,460	1,010,400	127,552	0	1,787,146
CORE SHROUD[a]	205,743	255,640	10,272,400	98,757	0	10,832,540
LOWER GRID PLATE[a]	221,445	319,550	1,684,000	106,294	0	2,331,289
LOWER SUPPORT COLUMN	56,181	63,910	168,400	26,967	0	315,458
LOWER CORE FORGING	610,397	703,010	1,052,500	292,990	0	2,658,897
MISC. INTERNALS	494,880	511,280	842,000	237,542	0	2,085,702
BIO SHIELD CONCRETE	0	0	0	0	2,571,846	2,571,846
REACTOR CAVITY LINER	242,944	0	4,210	0	0	247,154
REACTOR COOLANT PUMPS	0	0	0	0	991,810	991,810
PRESSURIZER	0	0	0	0	257,185	257,185
R.Hx, EHx, SUMP PUMP, CAVITY PUMP	0	0	0	0	15,563	15,563
PRESSURIZER RELIEF TANK	0	0	0	0	35,874	35,874
SAFETY INJECTION ACCUM. TANKS	0	0	0	0	403,582	403,582
STEAM GENERATORS	0	0	0	0	3,629,601	3,629,601
REACTOR COOLANT PIPING	0	0	0	0	292,792	292,792
REMAINING CONTAM. MATLS	0	0	0	0	5,176,006	5,176,006
CONTAM. MATL OTHER BLDG	0	0	0	0	39,819,187	39,819,187
FILTER CARTRIDGES	0	0	0	0	71,220	71,220
SPENT RESINS	1,113,480	1,278,200	3,368,000	534,470	0	6,294,150
COMBUSTIBLE WASTES	0	0	0	0	712,204	712,204
EVAPORATOR BOTTOMS	5,233,356	6,007,540	15,829,600	714,854	0	27,785,350
POST-TMI-2 ADDITIONS	10,501,290	0	0	0	0	10,501,290
SITE ACCESS FEES (3 YRS)						0
SUBTOTAL PWR COSTS	**26,863,286**	**17,894,800**	**47,295,140**	**4,786,297**	**53,976,869**	**150,816,392**
BARNWELL COUNTY BUSINESS TAX						0
ATLANTIC COMPACT SURCHARGE (INSIDE COMPACT)						3,883,482
TOTAL PWR COSTS (INSIDE COMPACT)						**154,699,874**

(a) GTCC Material: Assumes a low-density, distributed packaging scheme and final disposal as LLW. High-density packaging, ISFSI storage, and geologic repository disposal could reduce disposal costs.

Table B-46. BWR Disposition Costs Using Waste Vendors with Burial Costs at the South Carolina Site Atlantic Compact (2006 dollars)

REFERENCE BWR COMPONENT	BASE DISPOSAL CHARGE	CASK HANDLING	CURIE SURCHARGE	DOSE RATE SURCHARGE	WASTE VENDOR CHARGE	DISPOSAL COST
STEAM SEPARATOR	205,580	1,789,480	2,357,600	98,678	0	4,451,338
FUEL SUPPORT & PIECES	90,557	894,740	294,700	43,468	0	1,323,465
CONTROL RODS/INCORES	269,611	511,280	1,347,200	129,413	0	2,257,504
CONTROL RODS GUIDES	75,783	766,920	42,100	28,040	0	912,843
JET PUMPS	219,231	2,556,400	3,368,000	105,231	0	6,248,861
TOP FUEL GUIDES	375,571	4,601,520	12,124,800	180,274	0	17,282,165
CORE SUPPORT PLATE	251,770	1,981,210	273,650	93,155	0	2,599,785
CORE SHROUD(a)	735,197	8,947,400	23,576,000	352,895	0	33,611,492
REACTOR VESSEL WALL	159,949	1,406,020	909,360	59,181	0	2,534,509
SAC SHIELD	0	0	0	0	1,115,496	1,115,496
REACTOR WATER REC	0	0	0	0	357,532	357,532
SAC SHIELD	0	0	0	0	2,888,796	2,888,796
OTHER PRIMARY CONTAINMENT	0	0	0	0	11,577,329	11,577,329
CONTAINMENT ATMOSPHERIC	0	0	0	0	46,603	46,603
HIGH PRESSURE CORE SPRAY	0	0	0	0	113,712	113,712
LOW PRESSURE CORE SPRAY	0	0	0	0	41,011	41,011
REACTOR BLDG CLOSED COOLING	0	0	0	0	111,549	111,549
REACTOR CORE ISO COOLING	0	0	0	0	36,212	36,212
RESIDUAL HEAT REMOVAL	0	0	0	0	343,003	343,003
POOL LINER & RACKS	0	0	0	0	1,421,842	1,421,842
CONTAMINATED CONCRETE	0	0	0	0	2,032,902	2,032,902
OTHER REACTOR BUILDING	0	0	0	0	2,533,609	2,533,609
TURBINE	0	0	0	0	6,819,182	6,819,182
NUCLEAR STEAM CONDENSATE	0	0	0	0	901,293	901,293
LOW PRESSURE FEEDWATER HEATERS	0	0	0	0	3,026,592	3,026,592
MAIN STEAM	0	0	0	0	132,355	132,355
MOISTURE SEPARATOR REHEATERS	0	0	0	0	1,711,419	1,711,419
REACTOR FEEDWATER PUMPS	0	0	0	0	450,175	450,175
HIGH PRESSURE FEEDWATER HEATERS	0	0	0	0	602,694	602,694
OTHER TG BLDG	0	0	0	0	13,081,385	13,081,385
RAD WASTE BLDG	0	0	0	0	4,468,119	4,468,119
REACTOR BLDG	0	0	0	0	3,187,390	3,187,390
TG BLDG	0	0	0	0	2,096,967	2,096,967
RAD WASTE & CONTROL	0	0	0	0	1,929,210	1,929,210
CONCENTRATOR BOTTOMS	21,504,964	14,379,750	37,890,000	2,913,206	0	76,687,920
OTHER	5,830,235	3,898,510	403,739	145,278	0	10,277,762
POST-TMI-2 ADDITIONS	858,295	0	0	0	0	858,295
SITE ACCESS FEES (3.5 YRS)						0
SUBTOTAL BWR COSTS	**30,576,742**	**41,733,230**	**82,587,149**	**4,148,818**	**61,026,373**	**220,072,312**
BARNWELL COUNTY BUSINESS TAX						0
ATLANTIC COMPACT SURCHARGE (INSIDE COMPACT)						4,021,086
TOTAL BWR COSTS (INSIDE COMPACT)						**224,093,398**

(a) GTCC Material: Assumes a low-density, distributed packaging scheme and final disposal as LLW. High-density packaging, ISFSI storage, and geologic repository disposal could reduce disposal costs.

Table B-47. PWR Disposition Costs Using Waste Vendors with Burial Costs at the South Carolina Site Non-Atlantic Compact (2006 dollars)

REFERENCE PWR COMPONENT	BASE DISPOSAL CHARGE	CASK HANDLING	CURIE SURCHARGE	DOSE RATE SURCHARGE	WASTE VENDOR CHARGE	DISPOSAL COST
VESSEL WALL	3,207,162	2,686,790	6,874,200	1,539,438	0	14,307,589
VESSEL HEAD & BOTTOM	1,929,813	2,828,200	9,000	0	0	4,767,013
UPPER CORE SUPPORT ASSM	191,569	282,820	4,500	61,302	0	540,190
UPPER SUPPORT COLUMN	188,724	282,820	45,000	60,392	0	576,935
UPPER CORE BARREL	71,112	141,410	362,000	34,134	0	608,656
UPPER CORE GRID PLATE	163,200	353,525	943,380	78,336	0	1,538,441
GUIDE TUBES	288,112	424,230	45,000	77,790	0	835,132
LOWER CORE BARREL[a]	1,336,868	2,262,560	6,932,032	641,697	0	11,173,157
THERMAL SHIELDS[a]	258,980	424,230	1,340,000	124,310	0	2,147,520
CORE SHROUD[a]	195,906	282,820	11,381,968	94,035	0	11,954,729
LOWER GRID PLATE[a]	185,597	353,525	2,293,800	89,087	0	2,922,009
LOWER SUPPORT COLUMN	50,875	70,705	200,000	24,420	0	346,000
LOWER CORE FORGING	552,750	777,755	1,125,000	265,320	0	2,720,825
MISC. INTERNALS	455,120	565,640	900,000	218,458	0	2,139,218
BIO SHIELD CONCRETE	0	0	0	0	2,571,846	2,571,846
REACTOR CAVITY LINER	246,496	0	4,500	0	0	250,996
REACTOR COOLANT PUMPS	0	0	0	0	991,810	991,810
PRESSURIZER	0	0	0	0	257,185	257,185
R.Hx, EHx, SUMP PUMP, CAVITY PUMP	0	0	0	0	15,563	15,563
PRESSURIZER RELIEF TANK	0	0	0	0	35,874	35,874
SAFETY INJECTION ACCUM. TANKS	0	0	0	0	403,582	403,582
STEAM GENERATORS	0	0	0	0	3,629,601	3,629,601
REACTOR COOLANT PIPING	0	0	0	0	292,792	292,792
REMAINING CONTAM. MATLS	0	0	0	0	5,176,006	5,176,006
CONTAM. MATL OTHER BLDG	0	0	0	0	39,819,187	39,819,187
FILTER CARTRIDGES	0	0	0	0	71,220	71,220
SPENT RESINS	1,055,880	1,414,100	3,684,000	506,822	0	6,660,802
COMBUSTIBLE WASTES	0	0	0	0	712,204	712,204
EVAPORATOR BOTTOMS	4,962,636	6,646,270	17,000,528	487,817	0	29,097,251
POST-TMI-2 ADDITIONS	5,754,886	0	0	0	0	5,754,886
SITE ACCESS FEES (3 YRS)						0
SUBTOTAL PWR COSTS	**21,095,684**	**19,797,400**	**53,144,908**	**4,303,356**	**53,976,869**	**152,318,217**
BARNWELL COUNTY BUSINESS TAX						0
ATLANTIC COMPACT SURCHARGE (OUTSIDE COMPACT)						3,883,482
TOTAL PWR COSTS (OUTSIDE COMPACT)						**156,201,699**

(a) GTCC Material: Assumes a low-density, distributed packaging scheme and final disposal as LLW. High-density packaging, ISFSI storage, and geologic repository disposal could reduce disposal costs.

Table B-48. BWR Disposition Costs Using Waste Vendors with Burial Costs at the South Carolina Site Non-Atlantic Compact (2006 dollars)

REFERENCE BWR COMPONENT	BASE DISPOSAL CHARGE	CASK HANDLING	CURIE SURCHARGE	DOSE RATE SURCHARGE	WASTE VENDOR CHARGE	DISPOSAL COST
STEAM SEPARATOR	180,723	1,979,740	2,539,208	86,747	0	4,786,418
FUEL SUPPORT & PIECES	86,228	989,870	315,000	41,389	0	1,432,487
CONTROL RODS/INCORES	263,465	565,640	1,818,000	126,463	0	2,773,569
CONTROL RODS GUIDES	73,011	848,460	45,000	27,014	0	993,486
JET PUMPS	187,730	2,828,200	3,640,000	90,111	0	6,746,041
TOP FUEL GUIDES	331,666	5,090,760	13,020,192	159,200	0	18,601,818
CORE SUPPORT PLATE	231,542	2,191,855	292,500	85,671	0	2,801,568
CORE SHROUD[a]	751,170	9,898,700	37,800,000	360,562	0	48,810,432
REACTOR VESSEL WALL	144,843	1,555,510	972,000	53,592	0	2,725,945
SAC SHIELD	0	0	0	0	1,115,496	1,115,496
REACTOR WATER REC	0	0	0	0	357,532	357,532
SAC SHIELD	0	0	0	0	2,888,796	2,888,796
OTHER PRIMARY CONTAINMENT	0	0	0	0	11,577,329	11,577,329
CONTAINMENT ATMOSPHERIC	0	0	0	0	46,603	46,603
HIGH PRESSURE CORE SPRAY	0	0	0	0	113,712	113,712
LOW PRESSURE CORE SPRAY	0	0	0	0	41,011	41,011
REACTOR BLDG CLOSED COOLING	0	0	0	0	111,549	111,549
REACTOR CORE ISO COOLING	0	0	0	0	36,212	36,212
RESIDUAL HEAT REMOVAL	0	0	0	0	343,003	343,003
POOL LINER & RACKS	0	0	0	0	1,421,842	1,421,842
CONTAMINATED CONCRETE	0	0	0	0	2,032,902	2,032,902
OTHER REACTOR BUILDING	0	0	0	0	2,533,609	2,533,609
TURBINE	0	0	0	0	6,819,182	6,819,182
NUCLEAR STEAM CONDENSATE	0	0	0	0	901,293	901,293
LOW PRESSURE FEEDWATER HEATERS	0	0	0	0	3,026,592	3,026,592
MAIN STEAM	0	0	0	0	132,355	132,355
MOISTURE SEPARATOR REHEATERS	0	0	0	0	1,711,419	1,711,419
REACTOR FEEDWATER PUMPS	0	0	0	0	450,175	450,175
HIGH PRESSURE FEEDWATER HEATERS	0	0	0	0	602,694	602,694
OTHER TG BLDG	0	0	0	0	13,081,385	13,081,385
RAD WASTE BLDG	0	0	0	0	4,468,119	4,468,119
REACTOR BLDG	0	0	0	0	3,187,390	3,187,390
TG BLDG	0	0	0	0	2,096,967	2,096,967
RAD WASTE & CONTROL	0	0	0	0	1,929,210	1,929,210
CONCENTRATOR BOTTOMS	20,621,513	15,908,625	40,690,470	2,001,662	0	79,222,269
OTHER	5,590,721	4,313,005	431,550	0	0	10,335,276
POST-TMI-2 ADDITIONS	470,360	0	0	0	0	470,360
SITE ACCESS FEES (3.5 YRS)						0
SUBTOTAL BWR COSTS	**28,932,973**	**46,170,365**	**101,563,920**	**3,032,410**	**61,026,373**	**240,726,042**
BARNWELL COUNTY BUSINESS TAX						0
ATLANTIC COMPACT SURCHARGE (OUTSIDE COMPACT)						4,021,086
TOTAL BWR COSTS (OUTSIDE COMPACT)						**244,747,128**

(a) GTCC Material: Assumes a low-density, distributed packaging scheme and final disposal as LLW. High-density packaging, ISFSI storage, and geologic repository disposal could reduce disposal costs.

Table B-49. PWR Disposition Costs Using Waste Vendors with Burial Costs at the Washington Site (2004 dollars)

REFERENCE PWR COMPONENT	VOLUME CHARGE	SHIPMENT CHARGE	CONTAINER CHARGE	LINER DOSE RATE CHARGE	WASTE VENDOR CHARGE	DISPOSAL COST
VESSEL WALL	215,080	373,160	187,340	1,520,000	0	2,295,580
VESSEL HEAD & BOTTOM	226,400	392,800	197,200	3,800	0	820,200
UPPER CORE SUPPORT ASSM	22,640	39,280	19,720	107,200	0	188,840
UPPER SUPPORT COLUMN	22,640	39,280	19,720	107,200	0	188,840
UPPER CORE BARREL	11,320	19,640	9,860	80,000	0	120,820
UPPER CORE GRID PLATE	28,300	49,100	24,650	200,000	0	302,050
GUIDE TUBES	33,960	58,920	29,580	160,800	0	283,260
LOWER CORE BARREL[a]	181,120	314,240	157,760	1,280,000	0	1,933,120
THERMAL SHIELDS[a]	33,960	58,920	29,580	240,000	0	362,460
CORE SHROUD[a]	22,640	39,280	19,720	160,000	0	241,640
LOWER GRID PLATE[a]	28,300	49,100	24,650	200,000	0	302,050
LOWER SUPPORT COLUMN	5,660	9,820	4,930	40,000	0	60,410
LOWER CORE FORGING	62,260	108,020	54,230	440,000	0	664,510
MISC. INTERNALS	45,280	78,560	39,440	320,000	0	483,280
BIO SHIELD CONCRETE	0	0	0	0	2,571,846	2,571,846
REACTOR CAVITY LINER	28,979	9,820	19,720	0	0	58,519
REACTOR COOLANT PUMPS	0	0	0	0	991,810	991,810
PRESSURIZER	0	0	0	0	257,185	257,185
R.Hx, EHx, SUMP PUMP, CAVITY PUMP	0	0	0	0	15,563	15,563
PRESSURIZER RELIEF TANK	0	0	0	0	35,874	35,874
SAFETY INJECTION ACCUM. TANKS	0	0	0	0	403,582	403,582
STEAM GENERATORS	0	0	0	0	3,629,601	3,629,601
REACTOR COOLANT PIPING	0	0	0	0	292,792	292,792
REMAINING CONTAM. MATLS	0	0	0	0	5,176,006	5,176,006
CONTAM. MATL OTHER BLDG	0	0	0	0	39,819,187	39,819,187
FILTER CARTRIDGES	0	0	0	0	71,220	71,220
SPENT RESINS	113,200	196,400	98,600	800,000	0	1,208,200
COMBUSTIBLE WASTES	0	0	0	0	712,204	712,204
EVAPORATOR BOTTOMS	532,040	923,080	463,420	1,186,315	0	3,104,855
POST-TMI-2 ADDITIONS	880,866	0	0	0	0	880,866
HEAVY OBJECT SURCHARGE						0
SITE AVAILABILITY CHARGES (3 YRS)						382,821
SUBTOTAL PWR COSTS	**2,494,645**	**2,759,420**	**1,400,120**	**6,845,315**	**53,976,869**	**67,859,190**
TAXES & FEES (% OF CHARGES)						596,940
TAXES & FEES ($/CU.FT.)						599,569
ANNUAL PERMIT FEES (3 YRS)						127,200
TOTAL PWR COSTS						**69,182,899**

(a) GTCC Material: Assumes a low-density, distributed packaging scheme and final disposal as LLW. High-density packaging, ISFSI storage, and geologic repository disposal could reduce disposal costs.

Table B-50. BWR Disposition Costs Using Waste Vendors with Burial Costs at the Washington Site (2004 dollars)

REFERENCE BWR COMPONENT	VOLUME CHARGE	SHIPMENT CHARGE	CONTAINER CHARGE	LINER DOSE RATE CHARGE	WASTE VENDOR CHARGE	DISPOSAL COST
STEAM SEPARATOR	19,980	137,480	138,040	18,816,000	0	19,111,500
FUEL SUPPORT & PIECES	10,018	68,740	69,020	560,000	0	707,778
CONTROL RODS/INCORES	29,998	78,560	39,440	5,376,000	0	5,523,998
CONTROL RODS GUIDES	7,981	58,920	59,160	480,000	0	606,061
JET PUMPS	28,017	196,400	197,200	26,880,000	0	27,301,617
TOP FUEL GUIDES	47,997	707,040	354,960	48,384,000	0	49,493,997
CORE SUPPORT PLATE	22,017	157,120	152,830	1,240,000	0	1,571,967
CORE SHROUD[a]	93,956	1,374,800	690,200	94,080,000	0	96,238,956
REACTOR VESSEL WALL	16,018	196,400	108,460	880,000	0	1,200,878
SAC SHIELD (NEUTRON ACTIV. MATL)	0	0	0	0	1,115,496	1,115,496
REACTOR WATER REC	0	0	0	0	357,532	357,532
SAC SHIELD (CONTAM. MATL)	0	0	0	0	2,888,796	2,888,796
OTHER PRIMARY CONTAINMENT	0	0	0	0	11,577,329	11,577,329
CONTAINMENT ATMOSPHERIC	0	0	0	0	46,603	46,603
HIGH PRESSURE CORE SPRAY	0	0	0	0	113,712	113,712
LOW PRESSURE CORE SPRAY	0	0	0	0	41,011	41,011
REACTOR BLDG CLOSED COOLING	0	0	0	0	111,549	111,549
REACTOR CORE ISO COOLING	0	0	0	0	36,212	36,212
RESIDUAL HEAT REMOVAL	0	0	0	0	343,003	343,003
POOL LINER & RACKS	0	0	0	0	1,421,842	1,421,842
CONTAMINATED CONCRETE	0	0	0	0	2,032,902	2,032,902
OTHER REACTOR BUILDING	0	0	0	0	2,533,609	2,533,609
TURBINE	0	0	0	0	6,819,182	6,819,182
NUCLEAR STEAM CONDENSATE	0	0	0	0	901,293	901,293
LOW PRESSURE FEEDWATER HEATERS	0	0	0	0	3,026,592	3,026,592
MAIN STEAM	0	0	0	0	132,355	132,355
MOISTURE SEPARATOR REHEATERS	0	0	0	0	1,711,419	1,711,419
REACTOR FEEDWATER PUMPS	0	0	0	0	450,175	450,175
HIGH PRESSURE FEEDWATER HEATERS	0	0	0	0	602,694	602,694
OTHER TG BLDG	0	0	0	0	13,081,385	13,081,385
RAD WASTE BLDG	0	0	0	0	4,468,119	4,468,119
REACTOR BLDG	0	0	0	0	3,187,390	3,187,390
TG BLDG	0	0	0	0	2,096,967	2,096,967
RAD WASTE & CONTROL	0	0	0	0	1,929,210	1,929,210
CONCENTRATOR BOTTOMS	1,273,500	2,209,500	1,109,250	2,815,175	0	7,407,425
OTHER	345,260	599,020	300,730	132,240	0	1,377,250
POST-TMI-2 ADDITIONS	71,995	0	0	0	0	71,995
HEAVY OBJECT SURCHARGE						0
SITE AVAILABILITY CHARGES (3.5 YRS)						510,428
SUBTOTAL BWR COSTS	**1,966,737**	**5,783,980**	**3,219,290**	**199,643,415**	**61,026,373**	**272,150,223**
TAXES & FEES (% OF CHARGES)						9,078,326
TAXES & FEES ($/CU.FT.)						495,159
ANNUAL PERMIT FEES (3.5 YRS)						169,600
TOTAL BWR COSTS						**281,893,308**

(a) GTCC Material: Assumes a low-density, distributed packaging scheme and final disposal as LLW. High-density packaging, ISFSI storage, and geologic repository disposal could reduce disposal costs.

Table B-51. PWR Disposition Costs Using Waste Vendors with Burial Costs at the South Carolina Site Atlantic Compact (2004 dollars)

REFERENCE PWR COMPONENT	BASE DISPOSAL CHARGE	CASK HANDLING	CURIE SURCHARGE	DOSE RATE SURCHARGE	WASTE VENDOR CHARGE	DISPOSAL COST
VESSEL WALL	2,838,980	2,061,272	5,441,752	1,362,711	0	11,704,715
VESSEL HEAD & BOTTOM	1,808,550	2,169,760	7,160	0	0	3,985,470
UPPER CORE SUPPORT ASSM	170,740	216,976	3,580	54,637	0	445,932
UPPER SUPPORT COLUMN	157,854	216,976	35,800	50,513	0	461,143
UPPER CORE BARREL	75,177	108,488	286,408	36,085	0	506,158
UPPER CORE GRID PLATE	187,943	271,220	716,020	90,212	0	1,265,395
GUIDE TUBES	278,155	325,464	35,800	75,102	0	714,521
LOWER CORE BARREL[a]	1,202,832	1,735,808	4,582,528	577,359	0	8,098,527
THERMAL SHIELDS[a]	225,531	325,464	859,224	108,255	0	1,518,474
CORE SHROUD[a]	174,605	216,976	8,735,444	83,811	0	9,210,836
LOWER GRID PLATE[a]	187,943	271,220	1,432,040	90,212	0	1,981,415
LOWER SUPPORT COLUMN	47,678	54,244	143,204	22,886	0	268,012
LOWER CORE FORGING	518,017	596,684	895,000	248,648	0	2,258,349
MISC. INTERNALS	420,000	433,952	716,000	201,600	0	1,771,552
BIO SHIELD CONCRETE	0	0	0	0	2,571,846	2,571,846
REACTOR CAVITY LINER	206,176	0	3,580	0	0	209,756
REACTOR COOLANT PUMPS	0	0	0	0	991,810	991,810
PRESSURIZER	0	0	0	0	257,185	257,185
R.Hx, EHx, SUMP PUMP, CAVITY PUMP	0	0	0	0	15,563	15,563
PRESSURIZER RELIEF TANK	0	0	0	0	35,874	35,874
SAFETY INJECTION ACCUM. TANKS	0	0	0	0	403,582	403,582
STEAM GENERATORS	0	0	0	0	3,629,601	3,629,601
REACTOR COOLANT PIPING	0	0	0	0	292,792	292,792
REMAINING CONTAM. MATLS	0	0	0	0	5,176,006	5,176,006
CONTAM. MATL OTHER BLDG	0	0	0	0	39,819,187	39,819,187
FILTER CARTRIDGES	0	0	0	0	71,220	71,220
SPENT RESINS	945,000	1,084,880	2,864,080	453,600	0	5,347,560
COMBUSTIBLE WASTES	0	0	0	0	712,204	712,204
EVAPORATOR BOTTOMS	4,441,500	5,098,936	13,461,176	606,690	0	23,608,302
POST-TMI-2 ADDITIONS	8,913,864	0	0	0	0	8,913,864
SITE ACCESS FEES (3 YRS)						0
SUBTOTAL PWR COSTS	22,800,544	15,188,320	40,218,796	4,062,321	53,976,869	136,246,850
BARNWELL COUNTY BUSINESS TAX						0
ATLANTIC COMPACT SURCHARGE (INSIDE COMPACT)						3,883,482
TOTAL PWR COSTS (INSIDE COMPACT)						**140,130,332**

(a) GTCC Material: Assumes a low-density, distributed packaging scheme and final disposal as LLW. High-density packaging, ISFSI storage, and geologic repository disposal could reduce disposal costs.

Table B-52. BWR Disposition Costs Using Waste Vendors with Burial Costs at the South Carolina Site Atlantic Compact (2004 dollars)

REFERENCE BWR COMPONENT	BASE DISPOSAL CHARGE	CASK HANDLING	CURIE SURCHARGE	DOSE RATE SURCHARGE	WASTE VENDOR CHARGE	DISPOSAL COST
STEAM SEPARATOR	174,477	1,518,832	2,004,856	83,749	0	3,781,915
FUEL SUPPORT & PIECES	76,852	759,416	250,600	36,889	0	1,123,757
CONTROL RODS/INCORES	228,816	433,952	1,145,632	109,832	0	1,918,232
CONTROL RODS GUIDES	64,318	650,928	35,800	23,798	0	774,844
JET PUMPS	186,063	2,169,760	2,864,080	89,310	0	5,309,213
TOP FUEL GUIDES	318,750	3,905,568	10,310,688	153,000	0	14,688,007
CORE SUPPORT PLATE	213,675	1,681,564	232,700	79,060	0	2,206,999
CORE SHROUD[(a)]	623,969	7,594,160	20,048,560	299,505	0	28,566,194
REACTOR VESSEL WALL	135,741	1,193,368	773,280	50,224	0	2,152,614
SAC SHIELD (NEUTRON ACT. MATL)	0	0	0	0	1,115,496	1,115,496
REACTOR WATER REC	0	0	0	0	357,532	357,532
SAC SHIELD (CONTAM. MATL)	0	0	0	0	2,888,796	2,888,796
OTHER PRIMARY CONTAINMENT	0	0	0	0	11,577,329	11,577,329
CONTAINMENT ATMOSPHERIC	0	0	0	0	46,603	46,603
HIGH PRESSURE CORE SPRAY	0	0	0	0	113,712	113,712
LOW PRESSURE CORE SPRAY	0	0	0	0	41,011	41,011
REACTOR BLDG CLOSED COOLING	0	0	0	0	111,549	111,549
REACTOR CORE ISO COOLING	0	0	0	0	36,212	36,212
RESIDUAL HEAT REMOVAL	0	0	0	0	343,003	343,003
POOL LINER & RACKS	0	0	0	0	1,421,842	1,421,842
CONTAMINATED CONCRETE	0	0	0	0	2,032,902	2,032,902
OTHER REACTOR BUILDING	0	0	0	0	2,533,609	2,533,609
TURBINE	0	0	0	0	6,819,182	6,819,182
NUCLEAR STEAM CONDENSATE	0	0	0	0	901,293	901,293
LOW PRESSURE FEEDWATER HEATERS	0	0	0	0	3,026,592	3,026,592
MAIN STEAM	0	0	0	0	132,355	132,355
MOISTURE SEPARATOR REHEATERS	0	0	0	0	1,711,419	1,711,419
REACTOR FEEDWATER PUMPS	0	0	0	0	450,175	450,175
HIGH PRESSURE FEEDWATER HEATERS	0	0	0	0	602,694	602,694
OTHER TG BLDG	0	0	0	0	13,081,385	13,081,385
RAD WASTE BLDG	0	0	0	0	4,468,119	4,468,119
REACTOR BLDG	0	0	0	0	3,187,390	3,187,390
TG BLDG	0	0	0	0	2,096,967	2,096,967
RAD WASTE & CONTROL	0	0	0	0	1,929,210	1,929,210
CONCENTRATOR BOTTOMS	18,254,169	12,204,900	32,220,900	2,472,831	0	65,152,801
OTHER	4,948,908	3,308,884	343,322	123,317	0	8,724,431
POST-TMI-2 ADDITIONS	728,551	0	0	0	0	728,551
SITE ACCESS FEES (3.5 YRS)						0
SUBTOTAL BWR COSTS	**25,954,291**	**35,421,332**	**70,230,418**	**3,521,516**	**61,026,373**	**196,153,930**
BARNWELL COUNTY BUSINESS TAX						0
ATLANTIC COMPACT SURCHARGE (INSIDE COMPACT)						4,021,086
TOTAL BWR COSTS (INSIDE COMPACT)						**200,175,016**

(a) GTCC Material: Assumes a low-density, distributed packaging scheme and final disposal as LLW. High-density packaging, ISFSI storage, and geologic repository disposal could reduce disposal costs.

Table B-53. PWR Disposition Costs Using Waste Vendors with Burial Costs at the South Carolina Site Non-Atlantic Compact (2004 dollars)

REFERENCE PWR COMPONENT	BASE DISPOSAL CHARGE	CASK HANDLING	CURIE SURCHARGE	DOSE RATE SURCHARGE	WASTE VENDOR CHARGE	DISPOSAL COST
VESSEL WALL	2,841,954	2,380,320	6,064,800	1,364,138	0	12,651,213
VESSEL HEAD & BOTTOM	1,709,463	2,505,600	7,980	0	0	4,223,043
UPPER CORE SUPPORT ASSM	169,733	250,560	3,990	54,314	0	478,597
UPPER SUPPORT COLUMN	167,213	250,560	39,900	53,508	0	511,181
UPPER CORE BARREL	63,000	125,280	319,200	30,240	0	537,720
UPPER CORE GRID PLATE	144,585	313,200	798,000	69,401	0	1,325,186
GUIDE TUBES	255,245	375,840	39,900	68,916	0	739,901
LOWER CORE BARREL[a]	1,184,400	2,004,480	5,107,200	568,512	0	8,864,592
THERMAL SHIELDS[a]	229,425	375,840	957,600	110,124	0	1,672,989
CORE SHROUD[a]	173,576	250,560	9,735,600	83,316	0	10,243,052
LOWER GRID PLATE[a]	164,430	313,200	1,596,000	78,926	0	2,152,556
LOWER SUPPORT COLUMN	45,066	62,640	159,600	21,632	0	288,938
LOWER CORE FORGING	489,636	689,040	997,500	235,025	0	2,411,201
MISC. INTERNALS	403,200	501,120	798,000	193,536	0	1,895,856
BIO SHIELD CONCRETE	0	0	0	0	2,571,846	2,571,846
REACTOR CAVITY LINER	218,400	0	3,990	0	0	222,390
REACTOR COOLANT PUMPS	0	0	0	0	991,810	991,810
PRESSURIZER	0	0	0	0	257,185	257,185
R.Hx, EHx, SUMP PUMP, CAVITY PUMP	0	0	0	0	15,563	15,563
PRESSURIZER RELIEF TANK	0	0	0	0	35,874	35,874
SAFETY INJECTION ACCUM. TANKS	0	0	0	0	403,582	403,582
STEAM GENERATORS	0	0	0	0	3,629,601	3,629,601
REACTOR COOLANT PIPING	0	0	0	0	292,792	292,792
REMAINING CONTAM. MATLS	0	0	0	0	5,176,006	5,176,006
CONTAM. MATL OTHER BLDG	0	0	0	0	39,819,187	39,819,187
FILTER CARTRIDGES	0	0	0	0	71,220	71,220
SPENT RESINS	935,640	1,252,800	3,192,000	449,107	0	5,829,547
COMBUSTIBLE WASTES	0	0	0	0	712,204	712,204
EVAPORATOR BOTTOMS	4,397,508	5,888,160	15,002,400	432,266	0	25,720,334
POST-TMI-2 ADDITIONS	5,098,439	0	0	0	0	5,098,439
SITE ACCESS FEES (3 YRS)						0
SUBTOTAL PWR COSTS	**18,690,911**	**17,539,200**	**44,823,660**	**3,812,962**	**53,976,869**	**138,843,602**
BARNWELL COUNTY BUSINESS TAX						0
ATLANTIC COMPACT SURCHARGE (OUTSIDE COMPACT)						3,883,482
TOTAL PWR COSTS (OUTSIDE COMPACT)						**142,727,084**

(a) GTCC Material: Assumes a low-density, distributed packaging scheme and final disposal as LLW. High-density packaging, ISFSI storage, and geologic repository disposal could reduce disposal costs.

Disposal Cost Based on Flat Rate Calculation

Base Cost = (Waste Volume [ft^3]) * \$600/ft^3 = 42,075 * 600 =   25,245,000

Spent Resins = (Resin Volume [ft^3]) * \$1,800/ft^3 = 2000 * 1,800 =   3,600,000

Atlantic Compact Surcharge = Volume [ft^3] * \$6 ft^3 = 44,075 * 6 =   264,450

Vendor Costs   53,976,869

Total   **83,086,319**

Table B-54. BWR Disposition Costs Using Waste Vendors with Burial Costs at the South Carolina Site Non-Atlantic Compact (2004 dollars)

REFERENCE BWR COMPONENT	BASE DISPOSAL CHARGE	CASK HANDLING	CURIE SURCHARGE	DOSE RATE SURCHARGE	WASTE VENDOR CHARGE	DISPOSAL COST
STEAM SEPARATOR	160,107	1,753,920	2,234,400	76,851	0	4,225,278
FUEL SUPPORT & PIECES	76,399	876,960	279,300	36,671	0	1,269,330
CONTROL RODS/INCORES	233,392	501,120	1,276,800	112,028	0	2,123,341
CONTROL RODS GUIDES	64,680	751,680	39,900	23,932	0	880,192
JET PUMPS	166,320	2,505,600	3,192,000	79,834	0	5,943,754
TOP FUEL GUIDES	293,832	4,510,080	11,491,200	141,039	0	16,436,151
CORE SUPPORT PLATE	205,128	1,941,840	259,350	75,897	0	2,482,215
CORE SHROUD[a]	665,469	8,769,600	22,344,000	319,425	0	32,098,494
REACTOR VESSEL WALL	128,304	1,378,080	861,840	47,473	0	2,415,697
SAC SHIELD (NEUTRON ACT. MATL)	0	0	0	0	1,115,496	1,115,496
REACTOR WATER REC	0	0	0	0	357,532	357,532
SAC SHIELD (CONTAM. MATL)	0	0	0	0	2,888,796	2,888,796
OTHER PRIMARY CONTAINMENT	0	0	0	0	11,577,329	11,577,329
CONTAINMENT ATMOSPHERIC	0	0	0	0	46,603	46,603
HIGH PRESSURE CORE SPRAY	0	0	0	0	113,712	113,712
LOW PRESSURE CORE SPRAY	0	0	0	0	41,011	41,011
REACTOR BLDG CLOSED COOLING	0	0	0	0	111,549	111,549
REACTOR CORE ISO COOLING	0	0	0	0	36,212	36,212
RESIDUAL HEAT REMOVAL	0	0	0	0	343,003	343,003
POOL LINER & RACKS	0	0	0	0	1,421,842	1,421,842
CONTAMINATED CONCRETE	0	0	0	0	2,032,902	2,032,902
OTHER REACTOR BUILDING	0	0	0	0	2,533,609	2,533,609
TURBINE	0	0	0	0	6,819,182	6,819,182
NUCLEAR STEAM CONDENSATE	0	0	0	0	901,293	901,293
LOW PRESSURE FEEDWATER HEATERS	0	0	0	0	3,026,592	3,026,592
MAIN STEAM	0	0	0	0	132,355	132,355
MOISTURE SEPARATOR REHEATERS	0	0	0	0	1,711,419	1,711,419
REACTOR FEEDWATER PUMPS	0	0	0	0	450,175	450,175
HIGH PRESSURE FEEDWATER HEATERS	0	0	0	0	602,694	602,694
OTHER TG BLDG	0	0	0	0	13,081,385	13,081,385
RAD WASTE BLDG	0	0	0	0	4,468,119	4,468,119
REACTOR BLDG	0	0	0	0	3,187,390	3,187,390
TG BLDG	0	0	0	0	2,096,967	2,096,967
RAD WASTE & CONTROL	0	0	0	0	1,929,210	1,929,210
CONCENTRATOR BOTTOMS	18,273,292	14,094,000	35,910,000	1,773,728	0	70,051,019
OTHER	4,954,092	3,821,040	382,641	0	0	9,157,773
POST-TMI-2 ADDITIONS	416,707	0	0	0	0	416,707
SITE ACCESS FEES (3.5 YRS)						0
SUBTOTAL BWR COSTS	**25,637,722**	**40,903,920**	**78,271,431**	**2,686,878**	**61,026,373**	**208,526,325**
BARNWELL COUNTY BUSINESS TAX						0
ATLANTIC COMPACT SURCHARGE (OUTSIDE COMPACT)						4,021,086
TOTAL BWR COSTS (OUTSIDE COMPACT)						**212,547,411**

(a) GTCC Material: Assumes a low-density, distributed packaging scheme and final disposal as LLW. High-density packaging, ISFSI storage, and geologic repository disposal could reduce disposal costs.

Disposal Cost Based on Flat Rate Calculation

Base Cost = (Waste Volume [ft³]) * $600/ft³ = 34,748 * 600 =	20,848,800
Spent Resins = (Resin Volume [ft³]) * $1,800/ft³ = 0 * 1,800 =	0
Atlantic Compact Surcharge = Volume [ft³] * $6 ft³ = 34,748 * 6 =	208,488
Vendor Costs	61,026,373
Total	**82,083,661**

Table B-55. PWR Disposition Costs Using Waste Vendors with Burial Costs at the Washington Site (2002 dollars)

REFERENCE PWR COMPONENT	VOLUME CHARGE	SHIPMENT CHARGE	CONTAINER CHARGE	LINER DOSE RATE CHARGE	WASTE VENDOR CHARGE	DISPOSAL COST
VESSEL WALL	144,020	228,342	78,280	2,101,400	0	2,552,042
VESSEL HEAD & BOTTOM	151,600	240,360	82,400	5,200	0	479,560
UPPER CORE SUPPORT ASSM	15,160	24,036	8,240	147,200	0	194,636
UPPER SUPPORT COLUMN	15,160	24,036	8,240	147,200	0	194,636
UPPER CORE BARREL	7,580	12,018	4,120	110,600	0	134,318
UPPER CORE GRID PLATE	18,950	30,045	10,300	276,500	0	335,795
GUIDE TUBES	22,740	36,054	12,360	220,800	0	291,954
LOWER CORE BARREL[a]	121,280	192,288	65,920	1,769,600	0	2,149,088
THERMAL SHIELDS[a]	22,740	36,054	12,360	331,800	0	402,954
CORE SHROUD[a]	15,160	24,036	8,240	221,200	0	268,636
LOWER GRID PLATE[a]	18,950	30,045	10,300	276,500	0	335,795
LOWER SUPPORT COLUMN	3,790	6,009	2,060	55,300	0	67,159
LOWER CORE FORGING	41,690	66,099	22,660	608,300	0	738,749
MISC. INTERNALS	30,320	48,072	16,480	442,400	0	537,272
BIO SHIELD CONCRETE	0	0	0	0	4,210,923	4,210,923
REACTOR CAVITY LINER	19,405	6,009	8,240	0	0	33,654
REACTOR COOLANT PUMPS	0	0	0	0	1,623,905	1,623,905
PRESSURIZER	0	0	0	0	421,092	421,092
R.Hx, EHx, SUMP PUMP, CAVITY PUMP	0	0	0	0	25,481	25,481
PRESSURIZER RELIEF TANK	0	0	0	0	58,737	58,737
SAFETY INJECTION ACCUM. TANKS	0	0	0	0	660,791	660,791
STEAM GENERATORS	0	0	0	0	5,942,800	5,942,800
REACTOR COOLANT PIPING	0	0	0	0	479,393	479,393
REMAINING CONTAM. MATLS	0	0	0	0	8,474,753	8,474,753
CONTAM. MATL OTHER BLDG	0	0	0	0	65,196,558	65,196,558
FILTER CARTRIDGES	0	0	0	0	116,610	116,610
SPENT RESINS	75,800	120,180	41,200	1,106,000	0	1,343,180
COMBUSTIBLE WASTES	0	0	0	0	1,166,102	1,166,102
EVAPORATOR BOTTOMS	356,260	564,846	193,640	1,635,910	0	2,750,656
POST-TMI-2 ADDITIONS	589,838	0	0	0	0	589,838
HEAVY OBJECT SURCHARGE						0
SITE AVAILABILITY CHARGES (3 YRS)						372,474
SUBTOTAL PWR COSTS	1,670,443	1,688,529	585,040	9,455,910	88,377,147	102,149,542
TAXES & FEES (% OF CHARGES)						523,351
TAXES & FEES ($/UNIT VOL.)						599,569
ANNUAL PERMIT FEES (3 YRS)						123,300
TOTAL PWR COSTS						103,395,762

(a) GTCC Material: Assumes a low-density, distr buted packaging scheme and final disposal as LLW. High-density packaging, ISFSI storage, and geologic repository disposal could reduce disposal costs.

Table B-56. BWR Disposition Costs Using Waste Vendors with Burial Costs at the Washington Site (2002 dollars)

REFERENCE BWR COMPONENT	VOLUME CHARGE	SHIPMENT CHARGE	CONTAINER CHARGE	LINER DOSE RATE CHARGE	WASTE VENDOR CHARGE	DISPOSAL COST
STEAM SEPARATOR	13,379	84,126	57,680	25,984,000	0	26,139,185
FUEL SUPPORT & PIECES	6,708	42,063	28,840	774,200	0	851,811
CONTROL RODS/INCORES	20,087	48,072	16,480	7,424,000	0	7,508,639
CONTROL RODS GUIDES	5,344	36,054	24,720	663,600	0	729,718
JET PUMPS	18,761	120,180	82,400	37,120,000	0	37,341,341
TOP FUEL GUIDES	32,139	432,648	148,320	66,816,000	0	67,429,107
CORE SUPPORT PLATE	14,743	96,144	63,860	1,714,300	0	1,889,047
CORE SHROUD[a]	62,914	841,260	288,400	129,920,000	0	131,112,574
REACTOR VESSEL WALL	10,726	120,180	45,320	1,216,600	0	1,392,826
SAC SHIELD (NEUTRON ACT. MATL)	0	0	0	0	1,455,351	1,455,351
REACTOR WATER REC	0	0	0	0	466,460	466,460
SAC SHIELD (CONTAM. MATL)	0	0	0	0	3,768,918	3,768,918
OTHER PRIMARY CONTAINMENT	0	0	0	0	15,104,565	15,104,565
CONTAINMENT ATMOSPHERIC	0	0	0	0	60,802	60,802
HIGH PRESSURE CORE SPRAY	0	0	0	0	148,356	148,356
LOW PRESSURE CORE SPRAY	0	0	0	0	53,505	53,505
REACTOR BLDG CLOSED COOLING	0	0	0	0	145,535	145,535
REACTOR CORE ISO COOLING	0	0	0	0	47,245	47,245
RESIDUAL HEAT REMOVAL	0	0	0	0	447,504	447,504
POOL LINER & RACKS	0	0	0	0	1,855,031	1,855,031
CONTAMINATED CONCRETE	0	0	0	0	2,652,261	2,652,261
OTHER REACTOR BUILDING	0	0	0	0	3,305,518	3,305,518
TURBINE	0	0	0	0	8,896,765	8,896,765
NUCLEAR STEAM CONDENSATE	0	0	0	0	1,175,887	1,175,887
LOW PRESSURE FEEDWATER HEATERS	0	0	0	0	3,948,696	3,948,696
MAIN STEAM	0	0	0	0	172,679	172,679
MOISTURE SEPARATOR REHEATERS	0	0	0	0	2,232,832	2,232,832
REACTOR FEEDWATER PUMPS	0	0	0	0	587,328	587,328
HIGH PRESSURE FEEDWATER HEATERS	0	0	0	0	786,315	786,315
OTHER TG BLDG	0	0	0	0	17,066,857	17,066,857
RAD WASTE BLDG	0	0	0	0	5,829,410	5,829,410
REACTOR BLDG	0	0	0	0	4,158,484	4,158,484
TG BLDG	0	0	0	0	2,735,844	2,735,844
RAD WASTE & CONTROL	0	0	0	0	2,516,977	2,516,977
CONCENTRATOR BOTTOMS	852,750	1,352,025	463,500	3,881,970	0	6,550,245
OTHER	231,190	366,549	125,660	181,020	0	904,419
POST-TMI-2 ADDITIONS	48,209	0	0	0	0	48,209
HEAVY OBJECT SURCHARGE						0
SITE AVAILABILITY CHARGES (3.5 YRS)						496,632
SUBTOTAL BWR COSTS	**1,316,949**	**3,539,301**	**1,345,180**	**275,695,690**	**79,619,124**	**362,012,876**
TAXES & FEES (% OF CHARGES)						10,730,963
TAXES & FEES ($/UNIT VOL.)						495,159
ANNUAL PERMIT FEES (3.5 YRS)						164,400
TOTAL BWR COSTS						**373,403,397**

(a) GTCC Material: Assumes a low-density, distr buted packaging scheme and final disposal as LLW. High-density packaging, ISFSI storage, and geologic repository disposal could reduce disposal costs.

Table B-57. PWR Disposition Costs Using Waste Vendors with Burial Costs at the South Carolina Site Atlantic Compact (2002 dollars)

REFERENCE PWR COMPONENT	BASE DISPOSAL CHARGE	CASK HANDLING	CURIE SURCHARGE	DOSE RATE SURCHARGE	WASTE VENDOR CHARGE	DISPOSAL COST
VESSEL WALL	2,617,120	1,900,304	5,016,760	1,256,218	0	10,790,402
VESSEL HEAD & BOTTOM	1,667,358	2,000,320	6,600	0	0	3,674,278
UPPER CORE SUPPORT ASSM	157,410	200,032	3,300	50,371	0	411,113
UPPER SUPPORT COLUMN	145,530	200,032	33,000	46,570	0	425,132
UPPER CORE BARREL	69,300	100,016	264,040	33,264	0	466,620
UPPER CORE GRID PLATE	173,250	250,040	660,100	83,160	0	1,166,550
GUIDE TUBES	256,410	300,048	33,000	69,231	0	658,689
LOWER CORE BARREL[a]	1,108,800	1,600,256	4,224,640	532,224	0	7,465,920
THERMAL SHIELDS[a]	207,900	300,048	792,120	99,792	0	1,399,860
CORE SHROUD[a]	160,974	200,032	8,053,220	77,268	0	8,491,494
LOWER GRID PLATE[a]	173,250	250,040	1,320,200	83,160	0	1,826,650
LOWER SUPPORT COLUMN	43,956	50,008	132,020	21,099	0	247,083
LOWER CORE FORGING	477,576	550,088	825,000	229,236	0	2,081,900
MISC. INTERNALS	387,200	400,064	660,000	185,856	0	1,633,120
BIO SHIELD CONCRETE	0	0	0	0	4,210,923	4,210,923
REACTOR CAVITY LINER	190,080	0	3,300	0	0	193,380
REACTOR COOLANT PUMPS	0	0	0	0	1,623,905	1,623,905
PRESSURIZER	0	0	0	0	421,092	421,092
R.Hx, EHx, SUMP PUMP, CAVITY PUMP	0	0	0	0	25,481	25,481
PRESSURIZER RELIEF TANK	0	0	0	0	58,737	58,737
SAFETY INJECTION ACCUM. TANKS	0	0	0	0	660,791	660,791
STEAM GENERATORS	0	0	0	0	5,942,800	5,942,800
REACTOR COOLANT PIPING	0	0	0	0	479,393	479,393
REMAINING CONTAM. MATLS	0	0	0	0	8,474,753	8,474,753
CONTAM. MATL OTHER BLDG	0	0	0	0	65,196,558	65,196,558
FILTER CARTRIDGES	0	0	0	0	116,610	116,610
SPENT RESINS	871,200	1,000,160	2,640,400	418,176	0	4,929,936
COMBUSTIBLE WASTES	0	0	0	0	1,166,102	1,166,102
EVAPORATOR BOTTOMS	4,094,640	4,700,752	12,409,880	559,310	0	21,764,582
POST-TMI-2 ADDITIONS	8,217,949	0	0	0	0	8,217,949
SUBTOTAL PWR COSTS	**21,019,903**	**14,002,240**	**37,077,580**	**3,744,934**	**88,377,147**	**164,221,804**
ATLANTIC COMPACT SURCHARGE (INSIDE COMPACT)						2,588,988
TOTAL PWR COSTS (INSIDE COMPACT)						**166,810,792**

(a) GTCC Material: Assumes a low-density, distributed packaging scheme and final disposal as LLW. High-density packaging, ISFSI storage, and geologic repository disposal could reduce disposal costs.

Table B-58. BWR Disposition Costs Using Waste Vendors with Burial Costs at the South Carolina Site Atlantic Compact (2002 dollars)

REFERENCE BWR COMPONENT	BASE DISPOSAL CHARGE	CASK HANDLING	CURIE SURCHARGE	DOSE RATE SURCHARGE	WASTE VENDOR CHARGE	DISPOSAL COST
STEAM SEPARATOR	160,838	1,400,224	1,848,280	77,202	0	3,486,544
FUEL SUPPORT & PIECES	70,852	700,112	231,000	34,009	0	1,035,973
CONTROL RODS/INCORES	210,947	400,064	1,056,160	101,254	0	1,768,425
CONTROL RODS GUIDES	59,290	600,096	33,000	21,937	0	714,323
JET PUMPS	171,518	2,000,320	2,640,400	82,328	0	4,894,566
TOP FUEL GUIDES	293,832	3,600,576	9,505,440	141,039	0	13,540,887
CORE SUPPORT PLATE	196,988	1,550,248	214,500	72,886	0	2,034,622
CORE SHROUD[a]	575,190	7,001,120	18,482,800	276,091	0	26,335,201
REACTOR VESSEL WALL	125,144	1,100,176	712,800	46,303	0	1,984,423
SAC SHIELD (NEUTRON ACT. MATL)	0	0	0	0	1,455,351	1,455,351
REACTOR WATER REC	0	0	0	0	466,460	466,460
SAC SHIELD (CONTAM. MATL)	0	0	0	0	3,768,918	3,768,918
OTHER PRIMARY CONTAINMENT	0	0	0	0	15,104,565	15,104,565
CONTAINMENT ATMOSPHERIC	0	0	0	0	60,802	60,802
HIGH PRESSURE CORE SPRAY	0	0	0	0	148,356	148,356
LOW PRESSURE CORE SPRAY	0	0	0	0	53,505	53,505
REACTOR BLDG CLOSED COOLING	0	0	0	0	145,535	145,535
REACTOR CORE ISO COOLING	0	0	0	0	47,245	47,245
RESIDUAL HEAT REMOVAL	0	0	0	0	447,504	447,504
POOL LINER & RACKS	0	0	0	0	1,855,031	1,855,031
CONTAMINATED CONCRETE	0	0	0	0	2,652,261	2,652,261
OTHER REACTOR BUILDING	0	0	0	0	3,305,518	3,305,518
TURBINE	0	0	0	0	8,896,765	8,896,765
NUCLEAR STEAM CONDENSATE	0	0	0	0	1,175,887	1,175,887
LOW PRESSURE FEEDWATER HEATERS	0	0	0	0	3,948,696	3,948,696
MAIN STEAM	0	0	0	0	172,679	172,679
MOISTURE SEPARATOR REHEATERS	0	0	0	0	2,232,832	2,232,832
REACTOR FEEDWATER PUMPS	0	0	0	0	587,328	587,328
HIGH PRESSURE FEEDWATER HEATERS	0	0	0	0	786,315	786,315
OTHER TG BLDG	0	0	0	0	17,066,857	17,066,857
RAD WASTE BLDG	0	0	0	0	5,829,410	5,829,410
REACTOR BLDG	0	0	0	0	4,158,484	4,158,484
TG BLDG	0	0	0	0	2,735,844	2,735,844
RAD WASTE & CONTROL	0	0	0	0	2,516,977	2,516,977
CONCENTRATOR BOTTOMS	16,827,644	11,251,800	29,704,500	2,279,585	0	60,063,529
OTHER	4,562,161	3,050,488	316,470	113,680	0	8,042,799
POST-TMI-2 ADDITIONS	671,672	0	0	0	0	671,672
SUBTOTAL BWR COSTS	**23,926,075**	**32,655,224**	**64,745,350**	**3,246,316**	**79,619,124**	**204,192,089**
ATLANTIC COMPACT SURCHARGE (INSIDE COMPACT)						2,680,724
TOTAL BWR COSTS (INSIDE COMPACT)						**206,872,813**

(a) GTCC Material: Assumes a low-density, distr buted packaging scheme and final disposal as LLW. High-density packaging, ISFSI storage, and geologic repository disposal could reduce disposal costs.

Table B-59. PWR Disposition Costs Using Waste Vendors with Burial Costs at the South Carolina Site Non-Atlantic Compact (2002 dollars)

REFERENCE PWR COMPONENT	BASE DISPOSAL CHARGE	CASK HANDLING	CURIE SURCHARGE	DOSE RATE SURCHARGE	WASTE VENDOR CHARGE	DISPOSAL COST
VESSEL WALL	2,730,132	1,983,600	5,236,704	1,310,463	0	11,260,899
VESSEL HEAD & BOTTOM	1,740,340	2,088,000	7,600	0	0	3,835,940
UPPER CORE SUPPORT ASSM	164,300	208,800	3,800	52,576	0	429,476
UPPER SUPPORT COLUMN	151,900	208,800	38,000	48,608	0	447,308
UPPER CORE BARREL	72,360	104,400	275,616	34,733	0	487,109
UPPER CORE GRID PLATE	180,900	261,000	689,040	86,832	0	1,217,772
GUIDE TUBES	267,732	313,200	38,000	72,288	0	691,220
LOWER CORE BARREL[a]	1,157,760	1,670,400	4,409,856	555,725	0	7,793,741
THERMAL SHIELDS[a]	217,080	313,200	826,848	104,198	0	1,461,326
CORE SHROUD[a]	168,020	208,800	8,406,288	80,650	0	8,863,758
LOWER GRID PLATE[a]	180,900	261,000	1,378,080	86,832	0	1,906,812
LOWER SUPPORT COLUMN	45,880	52,200	137,808	22,022	0	257,910
LOWER CORE FORGING	498,480	574,200	950,000	239,270	0	2,261,950
MISC. INTERNALS	404,000	417,600	760,000	193,920	0	1,775,520
BIO SHIELD CONCRETE	0	0	0	0	4,210,923	4,210,923
REACTOR CAVITY LINER	198,400	0	3,800	0	0	202,200
REACTOR COOLANT PUMPS	0	0	0	0	1,623,905	1,623,905
PRESSURIZER	0	0	0	0	421,092	421,092
R.Hx, EHx, SUMP PUMP, CAVITY PUMP	0	0	0	0	25,481	25,481
PRESSURIZER RELIEF TANK	0	0	0	0	58,737	58,737
SAFETY INJECTION ACCUM. TANKS	0	0	0	0	660,791	660,791
STEAM GENERATORS	0	0	0	0	5,942,800	5,942,800
REACTOR COOLANT PIPING	0	0	0	0	479,393	479,393
REMAINING CONTAM. MATLS	0	0	0	0	8,474,753	8,474,753
CONTAM. MATL OTHER BLDG	0	0	0	0	65,196,558	65,196,558
FILTER CARTRIDGES	0	0	0	0	116,610	116,610
SPENT RESINS	909,000	1,044,000	2,756,160	436,320	0	5,145,480
COMBUSTIBLE WASTES	0	0	0	0	1,166,102	1,166,102
EVAPORATOR BOTTOMS	4,272,300	4,906,800	12,953,952	583,578	0	22,716,630
POST-TMI-2 ADDITIONS	8,572,815	0	0	0	0	8,572,815
SUBTOTAL PWR COSTS	**21,932,299**	**14,616,000**	**38,871,552**	**3,908,015**	**88,377,147**	**167,705,013**
ATLANTIC COMPACT SURCHARGE (OUTSIDE COMPACT)						2,588,988
TOTAL PWR COSTS (OUTSIDE COMPACT)						**170,294,001**

(a) GTCC Material: Assumes a low-density, distributed packaging scheme and final disposal as LLW. High-density packaging, ISFSI storage, and geologic repository disposal could reduce disposal costs.

NUREG-1307

Table B-60. BWR Disposition Costs Using Waste Vendors with Burial Costs at the South Carolina Site Non-Atlantic Compact (2002 dollars)

Chapter 2 REFERENCE BWR COMPONENT	BASE DISPOSAL CHARGE	CASK HANDLING	CURIE SURCHARGE	DOSE RATE SURCHARGE	WASTE VENDOR CHARGE	DISPOSAL COST
STEAM SEPARATOR	167,940	1,461,600	1,929,312	80,611	0	3,639,462
FUEL SUPPORT & PIECES	73,954	730,800	266,000	35,498	0	1,106,251
CONTROL RODS/INCORES	220,099	417,600	1,102,464	105,648	0	1,845,811
CONTROL RODS GUIDES	61,908	626,400	38,000	22,906	0	749,214
JET PUMPS	179,091	2,088,000	2,756,160	85,964	0	5,109,215
TOP FUEL GUIDES	306,806	3,758,400	9,922,176	147,267	0	14,134,649
CORE SUPPORT PLATE	205,535	1,618,200	247,000	76,048	0	2,146,783
CORE SHROUD[a]	600,588	7,308,000	19,293,120	288,282	0	27,489,990
REACTOR VESSEL WALL	130,622	1,148,400	820,800	48,330	0	2,148,152
SAC SHIELD (NEUTRON ACT. MATL)	0	0	0	0	1,455,351	1,455,351
REACT. WATER REC	0	0	0	0	466,460	466,460
SAC SHIELD (CONTAM. MATL)	0	0	0	0	3,768,918	3,768,918
OTHER PRIMARY CONTAINMENT	0	0	0	0	15,104,565	15,104,565
CONTAINMENT ATMOSPHERIC	0	0	0	0	60,802	60,802
HIGH PRESSURE CORE SPRAY	0	0	0	0	148,356	148,356
LOW PRESSURE CORE SPRAY	0	0	0	0	53,505	53,505
REACTOR BLDG CLOSED COOLING	0	0	0	0	145,535	145,535
REACTOR CORE ISO COOLING	0	0	0	0	47,245	47,245
RESIDUAL HEAT REMOVAL	0	0	0	0	447,504	447,504
POOL LINER & RACKS	0	0	0	0	1,855,031	1,855,031
CONTAMINATED CONCRETE	0	0	0	0	2,652,261	2,652,261
OTHER REACTOR BUILDING	0	0	0	0	3,305,518	3,305,518
TURBINE	0	0	0	0	8,896,765	8,896,765
NUCLEAR STEAM CONDENSATE	0	0	0	0	1,175,887	1,175,887
LOW PRESSURE FEEDWATER HEATERS	0	0	0	0	3,948,696	3,948,696
MAIN STEAM	0	0	0	0	172,679	172,679
MOISTURE SEPARATOR REHEATERS	0	0	0	0	2,232,832	2,232,832
REACTOR FEEDWATER PUMPS	0	0	0	0	587,328	587,328
HIGH PRESSURE FEEDWATER HEATERS	0	0	0	0	786,315	786,315
OTHER TG BLDG	0	0	0	0	17,066,857	17,066,857
RAD WASTE BLDG	0	0	0	0	5,829,410	5,829,410
REACTOR BLDG	0	0	0	0	4,158,484	4,158,484
TG BLDG	0	0	0	0	2,735,844	2,735,844
RAD WASTE & CONTROL	0	0	0	0	2,516,977	2,516,977
CONCENTRATOR BOTTOMS	17,554,292	11,745,000	31,006,800	2,378,021	0	62,684,114
OTHER	4,759,164	3,184,200	364,420	118,589	0	8,426,373
POST-TMI-2 ADDITIONS	700,676	0	0	0	0	700,676
SUBTOTAL BWR COSTS	**24,960,674**	**34,086,600**	**67,746,252**	**3,387,164**	**79,619,124**	**209,799,814**
ATLANTIC COMPACT SURCHARGE (OUTSIDE COMPACT)						2,680,724
TOTAL BWR COSTS (OUTSIDE COMPACT)						**212,480,538**

(a) GTCC Material: Assumes a low-density, distributed packaging scheme and final disposal as LLW. High-density packaging, ISFSI storage, and geologic repository disposal could reduce disposal costs.

APPENDIX C.

BUREAU OF LABOR STATISTICS ON THE INTERNET

APPENDIX C.

BUREAU OF LABOR STATISTICS ON THE INTERNET

For use in the adjustment formula in Chapter 3, the labor indexes for the first quarter of 2012 and the producer price indexes for March 2012 were obtained from the Bureau of Labor Statistics (BLS) data on the Internet.

These dates were chosen to agree, to the extent possible, with the effective dates of the waste burial rate schedules. Instructions for accessing and obtaining the specific indexes used in this report follow below.

Bureau of Labor Statistics Internet Data Page

To obtain reports of producer price indexes and labor indexes, proceed as follows:

1.   Enter the URL: **http://www.bls.gov/data/**

2.   Click on the item labeled *Series Report*.

3.   In the box labeled *Enter series id(s) below*, type in the following six series identifications (IDs), one ID per line:

Series ID	Producer Price Indexes
wpu0543	(Industrial electric power–used in calculation of P_x, per Section 3.2)
wpu0573	(Light fuel oils–used in calculation of F_x per Section 3.2)

Labor Indexes (Used in the calculation of L_x, per Section 3.1)

CIU201000000021OI	(Total compensation, private industry, Northeast region)
CIU201000000022OI	(Total compensation, private industry, South region)
CIU201000000023OI	(Total compensation, private industry, Midwest region)
CIU201000000024OI	(Total compensation, private industry, West region)

4.   Click the button labeled *Next*.

5.   In the box labeled *Year(s) to report for*, select the years you want.

6.   Click on the button labeled *Retrieve Data* and the six tables of data you requested will be displayed.

APPENDIX D.

REPRESENTATIVE EXAMPLES OF DECOMMISSIONING COSTS FOR 2002 THROUGH 2012

APPENDIX D.

REPRESENTATIVE EXAMPLES OF DECOMMISSIONING COSTS FOR 2002 THROUGH 2012

In Section 3.4 of this revision and the five previous revisions of NUREG-1307, decommissioning costs for four typical situations were developed. Results of these calculations are summarized below.

Example 1 (Compact-Affiliated Disposal Facility Only)

Reactor Type: Pressurized-Water Reactor (PWR)
Thermal Power Rating: 3400 Megawatt Thermal(MWth)
Location of Plant: Northwest Compact
LLW Burial Location: Washington

	2002	2004	2006	2008	2010	2012
L_x	1.775	1.984	2.11	2.23	2.29	2.38
E_x	0.985	1.483	2.152	2.746	2.139	2.704
B_x	3.634	5.374	6.829	8.283	8.035	7.335
Decommissioning Cost (Millions)	$219	$280	$331	$381	$371	$369

Example 2 (Compact-Affiliated Disposal Facility Only)

Reactor Type: PWR
Thermal Power Rating: 3400 MWth
Location of Plant: Atlantic Compact
LLW Burial Location: South Carolina (Atlantic Compact)

	2002	2004	2006	2008	2010	2012
L_x	1.862	2.070	2.21	2.33	2.41	2.52
E_x	0.985	1.483	2.152	2.746	2.139	2.704
B_x	17.922	19.500	22.933	25.231	27.292	30. 581
Decommissioning Cost (Millions)	$555	$612	$710	$779	$824	$915

Example 3 (Combination of Waste Vendor/Non-Compact and Compact-Affiliated Disposal)

Reactor Type: PWR
Thermal Power Rating: 3400 MWth
Location of Plant: Atlantic Compact
LLW Burial Location: South Carolina (Atlantic Compact)

	2002	2004	2006	2008	2010	2012
L_x	1.862	2.070	2.21	2.33	2.41	2.52
E_x	0.985	1.483	2.152	2.746	2.139	2.704
B_x	9.273	7.790	8.600	9.872	12.280	13.885
Decommissioning Cost (Millions)	$355	$341	$379	$425	$477	$530

Example 4 (Combination of Waste Vendor/Non-Compact and Compact-Affiliated Disposal)

Reactor Type: Boiling-Water Reactor
Thermal Power Rating: 3400 MWth
Location of Plant: Midwest Compact
LLW Burial Location: Before 2008–South Carolina (Non-Atlantic Compact), Beginning 2008–Unknown (Generic LLW Disposal Site)

	2002	2004	2006	2008	2010	2012
L_x	1.788	2.002	2.13	2.23	2.29	2.39
E_x	0.965	1.496	2.206	2.853	2.181	2.795
B_x	8.860	8.863	10.206	11.198	12.540	14.160
Decommissioning Cost (Millions)	$437	$465	$529	$578	$612	$679

APPENDIX E.

LOW-LEVEL WASTE COMPACTS

APPENDIX E. LOW-LEVEL WASTE COMPACTS

The figure and table below identify the composition of all LLW compacts as of May 2010 (source: NRC, http://www.nrc.gov/waste/llw-disposal/licensing/compacts.html).

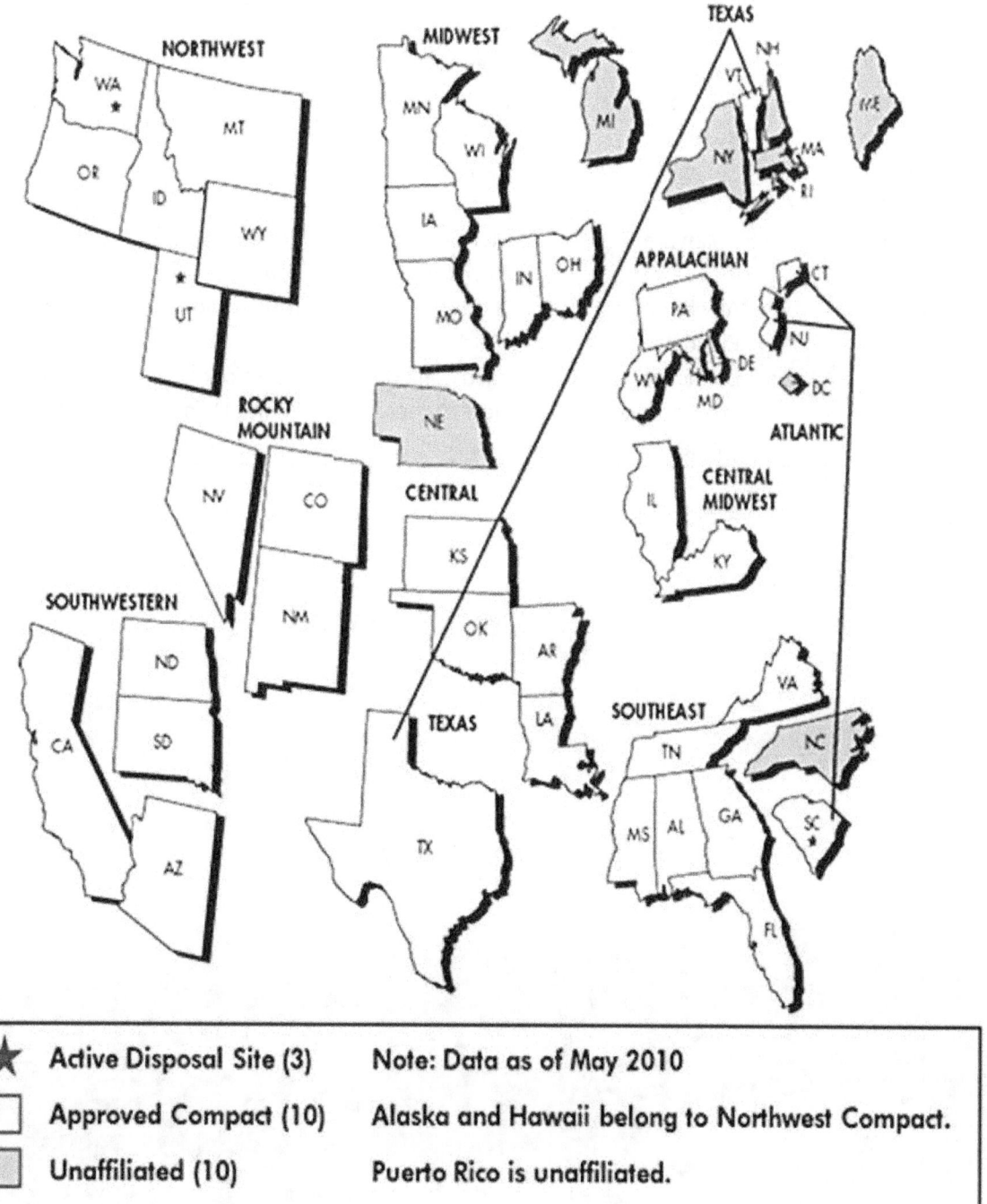

★	Active Disposal Site (3)	Note: Data as of May 2010
□	Approved Compact (10)	Alaska and Hawaii belong to Northwest Compact.
▨	Unaffiliated (10)	Puerto Rico is unaffiliated.

Compact	Affiliated States			
Northwest	Alaska	Idaho	Oregon	Washington[a]
	Hawaii	Montana	Utah	Wyoming
Southwestern	Arizona	California[b]	North Dakota	South Dakota
Rocky Mountain	Colorado	New Mexico	Nevada	
Midwest	Indiana	Minnesota	Ohio	Wisconsin
	Iowa	Missouri		
Central	Arkansas	Louisiana	Nebraska[b]	Oklahoma
	Kansas			
Texas	Texas[b]	Vermont		
Central Midwest	Illinois[b]	Kentucky		
Appalachian	Delaware	Maryland	Pennsylvania[b]	West Virginia
Atlantic	Connecticut[b]	New Jersey[b]	South Carolina[a]	
Southeast	Alabama	Georgia	Tennessee	Virginia
	Florida	Mississippi		
Unaffiliated States	District of Columbia	Michigan[b]	New York[b]	Rhode Island
	Massachusetts[b]	New Hampshire	Puerto Rico	North Carolina[b]
	Maine			

(a) Current Host State (2 States).
(b) Selected Host State (11 States).

NRC FORM 335
(12-2010)
NRCMD 3.7

U.S. NUCLEAR REGULATORY COMMISSION

BIBLIOGRAPHIC DATA SHEET

(See instructions on the reverse)

1. REPORT NUMBER (Assigned by NRC, Add Vol., Supp., Rev., and Addendum Numbers, If any.) NUREG-1307, Revision 15

2. TITLE AND SUBTITLE

NUREG-1307, Revision 15, "Report on Waste Burial Charges; Changes in Decommissioning Waste Disposal Costs at Low-Level Waste Burial Facilities"

Final Report

3. DATE REPORT PUBLISHED	
MONTH	YEAR
January	2013

4. FIN OR GRANT NUMBER

5. AUTHOR(S)

Pacific Northwest National Laboratory
Jason A. Gastelum and Steven Short

Jo Ann Simpson, Project Manager, USNRC

6. TYPE OF REPORT

Technical

7. PERIOD COVERED (Inclusive Dates)

8. PERFORMING ORGANIZATION - NAME AND ADDRESS (If NRC, provide Division, Office or Region, U. S. Nuclear Regulatory Commission, and mailing address; if contractor, provide name and mailing address.)

Pacific Northwest National Laboratory
902 Battelle Boulevard
P.O. Box 999
Richland, WA 99532

9. SPONSORING ORGANIZATION - NAME AND ADDRESS (If NRC, type "Same as above", if contractor, provide NRC Division, Office or Region, U. S. Nuclear Regulatory Commission, and mailing address.)

Division of Inspection and Regional Support
Office of Nuclear Reactor Regulation
U.S. Nuclear Regulatory Commission

10. SUPPLEMENTARY NOTES

Supercedes NUREG-1307, Revision 14, dated November 2010

11. ABSTRACT (200 words or less)

A requirement placed upon nuclear power reactor licensees by the U.S. Nuclear Regulatory Commission (NRC) is that licensees much annually adjust the estimate of the cost of decommissioning their power plants, in dollars of the current year, as part of the process to provide reasonable assurance that adequate funds for decommissioning will be available when needed. This report, which is revised periodically, explains the formula that is acceptable to the NRC for determining the minimum decommissioning funding assurance requirements for nuclear power plants. The sources of information used in the formula are identified, and the values developed fo rthe estimation of radioactive waste burial/disposition costs, by site and by year, are given. Licensees may use the formula, coefficients, and burial/disposition adjustment factors from this report in their cost analyses, or they may use adjustment factors derived from any methodology that results in a total cost estimate of no less than the amount estimated by using the parameters presented in this report.

12. KEY WORDS/DESCRIPTORS (List words or phrases that will assist researchers in locating the report.)

Waste burial
Decommissioning
Cost estimate
10 CFR 50.75(c)(2)

13. AVAILABILITY STATEMENT
unlimited
14. SECURITY CLASSIFICATION
(This Page)
unclassified
(This Report)
unclassified
15. NUMBER OF PAGES
16. PRICE

NUREG-1307, Rev. 15
Final

Report on Waste Burial Charges; Changes in Decommissioning
Waste Disposal Costs at Low-Level Waste Burial Facilities

January 2013

www.ingramcontent.com/pod-product-compliance
Lightning Source LLC
Chambersburg PA
CBHW080302180526
45167CB00006B/2630